# Nanotechnology Science and Technology

# Nanotechnology Science and Technology

**Magnetic Nanoparticles: Synthesis and Comprehensive Characterization**
V. N. Lad, PhD and Paritosh Agnihotri, PhD
2023. ISBN: 979-8-88697-961-9 (Softcover)
2023. ISBN: 979-8-89113-061-6 (eBook)

**Nanocomposite Hydrogels and Their Emerging Applications**
Mohammad Sirousazar, PhD **(Editor)**
2023. ISBN: 979-8-88697-675-5 (Hardcover)
2023. ISBN: 979-8-88697-895-7 (eBook)

**Applications of Gold Nanoparticles**
George L. Morrow (Editor)
2022. ISBN: 979-8-88697-272-6 (Hardcover)
2022. ISBN: 979-8-88697-301-3 (eBook)

**Titanium Dioxide: Advances in Research and Applications**
Aparna B. Gunjal, PhD (Editor)
2021. ISBN: 978-1-68507-457-9 (Softcover)
2021. ISBN: 978-1-68507-466-1 (eBook)

**Nano-Biotechnological Advancements in Environmental Issues: Applications and Challenges**
Ram Naresh Bharagava, PhD
and Reetika Singh, PhD (Editors)
2021. ISBN: 978-1-53619-975-8 (Hardcover)
2021. ISBN: 978-1-53619-984-0 (eBook)

More information about this series can be found at
https://novapublishers.com/product-category/series/nanotechnology-science-and-technology/

V. N. Lad
and Paritosh Agnihotri

# Magnetic Nanoparticles

Synthesis and Comprehensive Characterization

Copyright © 2023 by Nova Science Publishers, Inc.

DOI: https://doi.org/10.52305/EWET6012

**All rights reserved.** No part of this book may be reproduced, stored in a retrieval system or transmitted in any form or by any means: electronic, electrostatic, magnetic, tape, mechanical photocopying, recording or otherwise without the written permission of the Publisher.

We have partnered with Copyright Clearance Center to make it easy for you to obtain permissions to reuse content from this publication. Please visit copyright.com and search by Title, ISBN, or ISSN.

For further questions about using the service on copyright.com, please contact:

Copyright Clearance Center
Phone: +1-(978) 750-8400     Fax: +1-(978) 750-4470     E-mail: info@copyright.com

## NOTICE TO THE READER

The Publisher has taken reasonable care in the preparation of this book but makes no expressed or implied warranty of any kind and assumes no responsibility for any errors or omissions. No liability is assumed for incidental or consequential damages in connection with or arising out of information contained in this book. The Publisher shall not be liable for any special, consequential, or exemplary damages resulting, in whole or in part, from the readers' use of, or reliance upon, this material. Any parts of this book based on government reports are so indicated and copyright is claimed for those parts to the extent applicable to compilations of such works.

Independent verification should be sought for any data, advice or recommendations contained in this book. In addition, no responsibility is assumed by the Publisher for any injury and/or damage to persons or property arising from any methods, products, instructions, ideas or otherwise contained in this publication.

This publication is designed to provide accurate and authoritative information with regards to the subject matter covered herein. It is sold with the clear understanding that the Publisher is not engaged in rendering legal or any other professional services. If legal or any other expert assistance is required, the services of a competent person should be sought. FROM A DECLARATION OF PARTICIPANTS JOINTLY ADOPTED BY A COMMITTEE OF THE AMERICAN BAR ASSOCIATION AND A COMMITTEE OF PUBLISHERS.

## Library of Congress Cataloging-in-Publication Data

ISBN: 979-8-88697-961-9

Published by Nova Science Publishers, Inc. † New York

To Our Parents

*Jai Gurudev*

# Contents

**Preface** ................................................................................. xi

**Acknowledgments** ........................................................ xiii

**Chapter 1**     **Introduction** ............................................................. 1
                      1.1. Hematite ($\alpha$-$Fe_2O_3$) ................................................. 1
                      1.2. Magnetite ($Fe_3O_4$) ........................................................ 2
                      1.3. Maghemite ($\gamma$-$Fe_2O_3$) ........................................... 2
                      1.4. Iron Oxide Nanoparticles ............................................... 3

**Chapter 2**     **Basic Magnetic Properties** ................................. 5
                      2.1. Magnetism ..................................................................... 5
                             *2.1.1. Formation of Paramagnetism*
                                    *and Diamagnetism in Particles* ...................... 6
                      2.2. Magnetic Force .............................................................. 7
                             *2.2.1. Paramagnetism* ............................................ 8
                             *2.2.2. Diamagnetism* .............................................. 9
                             *2.2.3. Ferromagnetism* .......................................... 9
                             *2.2.4. Antiferromagnetism* .................................. 10
                             *2.2.5. Ferrimagnetism* ........................................ 10
                             *2.2.6. Superparamagnetism* ................................ 11
                      2.3. Coercivity .................................................................... 11
                      2.4. Coercivity and Remanence in Magnetic Particles ....... 12
                             *2.4.1. Terminology Involved in Magnetic*
                                    *Measurement* ................................................ 13
                      2.5. Magnetic Nanoparticles ............................................... 15
                      2.6. Magnetic Metal Oxide Nanoparticles (MMONPs) ..... 16
                      2.7. Magnetic Alloy Nanoparticles (MANPs) ................... 16

| Chapter 3 | **Synthesis of Magnetic Nanoparticles** ............................. 19 |
|---|---|
| | 3.1. Co-Precipitation ............................................................. 19 |
| | 3.2. Thermal Decomposition ................................................ 22 |
| | 3.3. Hydrothermal Synthesis ................................................ 24 |
| | 3.4. Microemulsion ............................................................... 25 |
| | 3.5. Sol-Gel Reaction Method .............................................. 27 |
| Chapter 4 | **Magnetic Properties of Magnetic Particles** .................... 29 |
| Chapter 5 | **Characterization of Particles** ........................................... 31 |
| | 5.1. Techniques of Characterization ..................................... 31 |
| | 5.2. Magnetism and Microfluidics ........................................ 32 |
| | 5.3. Characterization of Magnetic Nanoparticles ................. 32 |
| | 5.4. TEM and SEM Analysis ................................................ 33 |
| | 5.5. Size Dependent Characteristics of Iron Oxide Nanoparticles ............................................. 34 |
| | 5.6. Shape-Dependent Characteristics Iron Oxide Nanoparticles ............................................. 34 |
| | 5.7. Effects of Agglomeration of Magnetic Nanoparticles ............................................ 35 |
| | 5.8. Colloidal Stability of Magnetic Nanoparticles ............. 35 |
| |     *5.8.1. Surface Coatings of Nanoparticles ................. 36* |
| Chapter 6 | **Applications of Magnetic Nanoparticles** ......................... 53 |
| | 6.1. Heat Transfer Applications ............................................ 53 |
| | 6.2. Candidate for Next Generation Nuclear Waste Remediation Material ........................................ 54 |
| | 6.3. Magnetic Nanoparticles for Wastewater Treatment ...................................................................... 54 |
| | 6.4. Cooling in Nuclear Power Plants ................................... 54 |
| | 6.5. Applications in the Space and Defence Sectors ........... 55 |
| | 6.6. Other Thermal Applications .......................................... 55 |
| | 6.7. Cooling of Electronics Microchips ............................... 56 |
| | 6.8. Microfluidics Applications ............................................ 56 |
| | 6.9. Mechanical Applications ............................................... 57 |
| | 6.10. Magnetic Sealing ......................................................... 57 |
| | 6.11. Biomedical Applications ............................................. 58 |
| | 6.12. Magnetic Drug Delivery ............................................. 59 |
| | 6.13. Hyperthermia Therapy ................................................ 59 |

| | | |
|---|---|---|
| **Chapter 7** | **Concluding Remarks and Perspectives** | 61 |
| **References** | | 63 |
| | Further Reading | 80 |
| **Index** | | 81 |
| **About the Authors** | | 83 |

# Preface

The purpose of writing of this book is to provide basic information related to synthesis of magnetic nanoparticles, features of several synthesis methods, specific characterization, their stability and applications in different fields. The current book is applicable for students, researchers, academicians and working professionals of both science and engineering streams. It will find breakthrough in the scientific knowledge of current complex methods of nanoparticle synthesis. This book should help develop a strong intuitive sense for choosing methodology for magnetic nanoparticle synthesis and describing the formal methods. This is an introductory book which is applicable for teaching as well as for research purposes. It covers all the major aspects of nanoparticle synthesis in an easy and accessible ways. The initial chapters describe the types of magnetic particles, and magnetism in nanoparticles with their occurrences and behavior in different conditions. Later chapters describe the classical methods of nanoparticles synthesis, specific characterizations, stability, and finally broad applications in various technical fields in an exhaustive tabular format for easy and quick access for the readers.

It is assured that all the learnings help gain valuable knowledge from this book, and propagate with the basic quarries. We hope this book will be beneficial for the universities and institutes throughout the globe involved in the teaching-learning-research activities in the area of magnetic materials.

*V. N. Lad, PhD*
*Paritosh Agnihotri, PhD*

# Acknowledgments

Authors are thankful to Sardar Vallabhbhai National Institute of Technology – Surat, Gujarat, India for providing necessary support including infrastructure and research facilities. Authors also acknowledge the Ministry of Education, Government of India, for providing the scholarship to Paritosh Agnihotri during his doctoral work. The motivation and guidance provided by all the colleagues and faculty members of the department are highly appreciated. Acknowledgements are due to the support and cooperation extended by the members of the research group - Swati Ralekar, Prashant Deulgaonkar, Chirag Barasara, Stuti Dubey, Sandeep Joshi, Ridhdhi Tala, Priyanka Doot, Santosh Barik, Suhas Doke, Balaji Dhopte, Svapnil Kevat, Pavitra Sarang and Dr. C. M. Patel during the research work and/or during several writing phases and review phases of the book. Thanks are also due to Dr. Ankit Shah and Dr. N. I. Malek for their cooperation during the early stage of the research work.

We wish to express gratitude towards our family members for their moral support, motivation, understanding, and their patience during the preparation of the manuscript. Finally, we convey our sincere thanks to the publisher, Nova Publisher, New York, USA for the flexible processing of manuscript, both during editorial and production stages.

*V. N. Lad, PhD*
*Paritosh Agnihotri, PhD*

# Chapter 1

# Introduction

Iron and iron oxides with reasonable stability are present in very abundant amount in the nature which are useful for synthesis of versatile magnetic iron oxide nanoparticles [1]. Iron oxide nanoparticles exhibit specific magnetic and physiochemical properties that can be altered based on their size and shape depending on their end uses. Surface coated magnetic oxide nanoparticles can easily associate with biological species such as DNA, peptides and antibodies to form complexes useful for biomedical applications. Moreover, it is also possible to generate versatile nano-bio hybrid particles which simultaneously acquire magnetic and biological entities for diagnostics and therapeutics [2]. The controlled size and enhanced surface properties of these hybrid nanoparticles make them promising candidates to be used in variety of applications such as tracer probes and drug delivery vehicles for next generation diagnostic and therapeutic activities [3].

The oxides of iron are found in different forms such as FeO, $Fe_2O_3$ and $Fe_3O_4$ in the nature. $Fe_2O_3$ has four crystallographic phases namely, hematite ($\alpha$-$Fe_2O_3$), $\beta$-$Fe_2O_3$, $\varepsilon$-$Fe_2O_3$ and maghemite ($\gamma$-$Fe_2O_3$) [4, 5, 6, 7]. Sakurai et al. [8] reported the phase transformation of $Fe_2O_3$ phases as under for $\alpha, \gamma, \varepsilon, \beta,$ -phases.

$$\alpha \longrightarrow \gamma \longrightarrow \varepsilon \longrightarrow \beta$$

Magnetite and $\gamma$-$Fe_2O_3$ are very promising and widely acceptable for technical and industrial applications due to their biocompatibility, specific characteristics and other useful properties.

## 1.1. Hematite ($\alpha$-$Fe_2O_3$)

Hematite is one of the n-type semiconductors, and highly stable under ambient conditions. Hematite is used in many applications including catalysts, data storage, pigments, solar cells, water splitting and gas sensors due to its nontoxic, biodegradable and less corrosive character. [9, 10, 11, 12, 13]. Hematite is the working material for formation of magnetite and maghemite [14] which have been attracted by many scientists and researchers working in diversed fields since last few decades [15]. $Fe_2O_3$ is available in four different crystalline structures which are represented as $\gamma$-$Fe_2O_3$, $\beta$-$Fe_2O_3$, $\varepsilon$-$Fe_2O_3$ and $\alpha$-$Fe_2O_3$, out of which, $\gamma$-$Fe_2O_3$ and $\alpha$-$Fe_2O_3$ phases have been extensively studied and used for commercial purpose [4, 5, 16]. $\alpha$-$Fe_2O_3$ crystallizes in corundum structure which is having high thermodynamic stability. However, it shows week ferromagnetism at room temperature due to the Dzyaloshinsky–Moriya mechanism, and behaves like antiferromagnetism below -10°C [17, 18, 19, 20]. $\alpha$-$Fe_2O_3$ shows paramagnetism at temperature more than 677 °C (its Curie temperature).

## 1.2. Magnetite ($Fe_3O_4$)

The most common type of iron oxide is magnetite. It has face cantered cubic inverse spinel structure, with oxygen forming closed packing structure along the [111] direction. Magnetite has a cubic spinel structure. Hence, the electrons are occupied between $Fe^{2+}$ and $Fe^{3+}$ in the octahedral and tetrahedral sites at atmospheric temperature which makes magnetite as a classical material for magnetic applications [21, 22]. The divalent ions with the stoichiometric proportion $Fe^{+2}:Fe^{+3} = 1:2$ can be completely replaced by other divalent ions such as Co, Mn, Zn, etc. $Fe_3O_4$ shows the properties of n- and p-type semiconductors. Magnetite is preferred to other iron oxides due to its excellent magnetic property viz. high saturation magnetisation (84 emu $g^{-1}$) at room temperature [23]. Its magnetization and transportation are performed by well established Verwey (order–disorder) transition at -150°C which strongly affects the magnetisation, conductivity and coercivity of materials [24].

## 1.3. Maghemite ($\gamma$-$Fe_2O_3$)

The structure of $\gamma$-$Fe_2O_3$ is same as that of $\alpha$-$Fe_2O_3$ with the only difference from $Fe_3O_4$ is that all of Fe is in the trivalent state. $\gamma$-$Fe_2O_3$ is a low-cost and chemically stable ferrimagnetic material at room temperature which shows cubic spinel structure at around 928 K [5, 16]. Basically, $Fe_3O_4$ works as a precursor for synthesizing the γ-$Fe_2O_3$ which quickly transforms into α-$Fe_2O_3$ [3]. It is clearly reported [25, 26] that $Fe_3O_4$ to γ-$Fe_2O_3$ transformation at an elevated temperature near 200°C is usual due to their similar crystal structures. Since $Fe_3O_4$ structure is unstable, quick oxidation converts it into γ-$Fe_2O_3$. Moreover, conversion from γ-$Fe_2O_3$ to α-$Fe_2O_3$ requires very high temperature due to their completely different crystal structures. However, these conversions are irreversible at room temperature. The conversion of α-$Fe_2O_3$ to $Fe_3O_4$ is also possible at annealing temperature under reducing conditions of the surrounding medium. At the same instant, they can be converted into biocompatible materials exhibiting low toxicity [27, 28]. The chemical structure studies incorporating electron diffraction on HRTEM help differentiation between maghemite and magnetite.

The phase transformation of magnetite, maghemite and hematite is illustrated in [3]. These transformations are irreversible at atmospheric conditions [7]. The initial transformation from $Fe_3O_4$ to γ-$Fe_2O_3$ can be obtained at moderate temperate of 200°C because of their similar crystal structures [3]. Further, the phase transformation from γ-$Fe_2O_3$ to α-$Fe_2O_3$ obtained only at elevated temperature more than 500°C. As mentioned earlier, the crystal structures for this transformation are completely different.

## 1.4. Iron Oxide Nanoparticles

Iron oxide nanoparticles have different structures which constitute a superior category of functional materials useful in numerous fields [29, 30, 31, 32, 33, 34]. The synthesis of γ-$Fe_2O_3$ nanoparticles usually carried out by the calcination of iron hydroxide [35, 36] or by the oxidation of the prepared $Fe_3O_4$ nanoparticles, iron or organometallic iron [37, 38, 39]. The synthesis of α-$Fe_2O_3$ particles utilizes the acid hydrolysis of $FeCl_3$ at low pH [6, 29, 40, 41] that controls particle size, morphology, size distribution and other related properties [42, 43]. The synthesis of $Fe_3O_4$ particles is based on the reactions occur with the stoichiometric ratios of $Fe^{2+}$ and $Fe^{3+}$ in their aqueous solutions under different conditions [31, 44, 45]. However, these synthesis steps do not produce iron oxide nanoparticles with ultra-fine and narrow size distribution. The synthesis steps are very complex to control of nucleation and the particle growth. The scale-up procedure usually provides non-homogeneity in composition of solution during the reactions which also affects nucleation and polydispersity. Less quantity of iron oxide nanoparticles with narrow size distribution has been reported [46] due to this reason. A short nucleation time is required to achieve uniform size distribution of particles at the completion of reactions.

# Chapter 2

# Basic Magnetic Properties

## 2.1. Magnetism

Magnetic fields can be generated by electric current or by magnetic materials such as permanent magnets. The field of electromagnetism from these two different sources of magnetic field $\vec{B}$ are related as [47]

$$\vec{B} = \mu_0(\vec{H} + \vec{M}) \tag{1}$$

where $\vec{H}$ is the field generated as a result of externally controllable current in the coil of electromagnet. Magnetic fields produced by magnetic materials are the result of the well-known magnetization $\vec{M}$ (magnetic movement per volume) inside the material, whereas $\mu_0$ is the magnetic permeability for vacuum [52]. The magnetic particles may be classified based on their behaviour to external magnetic field.

There are different types of magnetism such as paramagnetism, diamagnetism, ferromagnetism, ferrimagnetism and antiferromagnetism [47]. The magnetic properties of the magnetic materials are based on the electronic structure of the atoms within the materials. There is nearly no resultant magnetic moment in most of the materials due to the electrons being coupled in anti-coupled pairs causing the magnetic moments to be repelled, especially in case of diamagnetism and paramagnetism. Diamagnetism generates from the orbital motion of electrons about the nuclei, and electromagnetically induced by the application of an external magnetic field. This type of magnetism is very weak, and it can be easily controlled by paramagnetism of atoms. This behaviour generates from magnetic atoms, and their rotations are isolated from their magnetic environment [25]. This nature of magnetism is also relatively weak, and thus materials exhibiting only diamagnetism or paramagnetism materials are basically known as non-magnetic materials. The conventional ferromagnetic materials have magnetic moments due to electrons responsible for exhibition of macroscopic magnetization. Magnetic material can also be ferrimagnetic which is basically found in various forms of compounds such as mixed oxides and ferrites. They do not only possess diamagnetic properties but also exhibit strong or permanent (ferromagnetic) magnetic properties. Magnetic susceptibility ($\chi$) is the measure of magnetic properties of any magnetic materials which is defined as

$$\vec{M} = \chi \vec{H} \tag{2}$$

where, $M$ is the magnetization and $H$ is the magnetic field.

## 2.1.1. Formation of Paramagnetism and Diamagnetism in Particles

Let us consider paramagnet and diamagnet under the influence of ambient temperature. The field strength in which magnetization is induced is linear with applied magnetic field as given by equation (1). The sign convention of $\chi$ in equation (2) identifies weather a material is paramagnetic or diamagnetic: If $\chi < 0$, material is diamagnetic, whereas for $\chi > 0$, the material is paramagnetic [48]. It means that the direction of $\vec{M}$ is parallel to $\vec{H}$ for paramagnetic materials, and opposite to $\vec{H}$ for diamagnetic materials. It is important to remember that the magnetization is induced by external field, and decrease to zero when the field is removed which is known as superparamagnetism [48].

By combining equations (1) and (2),

$$\vec{B} = \mu \vec{H} \tag{3}$$

where, $\mu \equiv \mu_0 (1 + \chi)$ being the magnetic permeability of the material. As mentioned earlier, for paramagnetic materials $\mu > \mu_0$ whereas for diamagnetic materials $\mu < \mu_0$. Generally, paramagnetism and diamagnetism are weak and typically $|\chi| \ll 1$, which shows that the effect of $\vec{M}$ on the total field is small: $\vec{B} \approx \mu_0 \vec{H}$. However, it is quite interesting to study the effects when a paramagnetic or diamagnetic object gets placed in an external field [52].

## 2.2. Magnetic Force

The magnetic energy $E$ of any object with volume $V$ is given by

$$E = -\frac{1}{2\mu_0}\chi V B^2 \qquad (4)$$

This equation implies that the energy of paramagnetic materials decreases with magnetic field, whereas it increases with magnetic field for diamagnetic materials [52]. Moreover, magnetic force and energy are related to each other as mentioned in eq. 5(a).

$$\vec{F} = -\vec{\nabla} E \qquad (5a)$$

The forces acting on the magnetic particles in presence of magnetic field depends on the volume of particles, the change in magnetic susceptibilities between the particles and surrounding buffer medium, along with the strength and gradient of the applied magnetic field can be mentioned as under.

$$\vec{F} = \frac{V \cdot \chi}{\mu_0}(\vec{B}.\vec{\nabla})\vec{B} \qquad (5b)$$

or

$$\vec{F} = \frac{V \cdot \Delta\chi}{\mu_0}(\vec{B}.\vec{\nabla})\vec{B} \qquad (5c)$$

where, $V$ is the volume of the particle $\Delta\chi$ difference in magnetic susceptibilities of particles and surrounding medium that is ($\chi_p - \chi_m$), $B$ is the magnetic induction, and $\mu_0 = 4\pi \times 10^{-7}(NA^{-2})$ is the permeability of the vacuum.

The above equation is applied in the case of homogenous magnetic field or $\nabla.B = 0$ with no force on the particles. Under this situation the particle must slightly be magnetised in the field but not to be dragged into any direction.

Pamme [49] described that the term $(\chi_p - \chi_m)$ is the difference in magnetic susceptibilities of particles $(\chi_p)$ and that of the surrounding medium

($\chi_m$) or buffer. For the diamagnetic substance ($\chi_p < 0$) in diamagnetic medium ($\chi_m < 0$) the value of $\Delta\chi$ can be positive or negative. It shows that particles can be diverted from or attached to the magnetic field. It may be noted that the magnetic susceptibilities of particles and surrounding medium is generally similar to each other so that the term $\Delta\chi$ is very small, the force on the particles is insignificant. If a diamagnetic substance ($\chi_p < 0$) is located in paramagnetic medium ($\chi_m > 0$) then the value of $\Delta\chi$ is always negative, and therefore diamagnetic substance will be migrated from the magnetic field of high strength. The higher $\chi_m$ situation, a paramagnetic substance ($\chi_p > 0$) can behave as a diamagnetic material in the presence of strong paramagnetic medium ($\chi_m > \chi_p > 0$). In this case $\Delta\chi$ is also negative, and therefore the paramagnetic substances are migrated from magnetic field. The above equations can be applied for paramagnetic and superparamagnetic substances. Singamaneni et al. [50] studied the properties of magnetic nanoparticles, and stated that without application of magnetic field, the average magnetization of the particle becomes zero and the diameter for particle goes below critical diameter, and so it becomes superparamagnetic in nature for which equation (3) is applicable.

Magnetic field lines can extend with specific density within the substance or material, and quantified by the material permeability (μ). The magnetic flux density ($B$) measured in Tesla (T) which correlates the number of field lines per unit area. The flux density reduces rapidly with increasing distance from the magnet surface [49]. If a magnetic material is placed in the magnetic field, the magnetic field lines deflect towards the material due to the influence of its higher permeability. Permanent magnets sustain their magnetic properties even after the removal of the external magnetic field. The materials that show this nature include iron, nickel and cobalt. As reported in [49], the strongest magnetic field can be achieved with alloys such as samarium cobalt (SmCo) or neodymium iron boron (NdFeB).

### 2.2.1. Paramagnetism

Paramagnetism is generated by unpaired pairs of electrons and atoms, and influenced by the external magnetic field. The thermal fluctuations may cause the magnetic movement of the atoms to move randomly in the absence of magnetic field.

## 2.2.2. Diamagnetism

Diamagnetism is the basic and extremely weak property of the materials. The magnetic susceptibility $(\chi)$ is negative and in the order of $10^{-5}$ which is independent of temperature. The diamagnetism generates from noncooperative nature of orbital electrons under the presence of an applied external magnetic field. Diamagnetic materials are confined with atoms which do not posses net magnetic movements. It means that all the orbital shells are filled, and no unpaired electrons are available. A negative magnetization is developed in such situation which is proportional to the magnetic field strength.

## 2.2.3. Ferromagnetism

Ferromagnetism is characterized by the parallel orientation of magnetic moments of molecules to one another which promotes higher magnetization even in the absence of magnetic field, also known as spontaneous magnetization. The atomic dipole moment is formed due to higher positive interactions generated by electronic forces resulted in parallel alignment of the atomic molecules. Magnetic susceptible temperature and spontaneous magnetization are two classical properties of the ferromagnetic material. Ferromagnetism is basically depends on temperature, and magnetization is inversely proportional to temperature as given by the Curie-Weiss law represented by equation (6).

$$\chi = \frac{C}{T-\theta} \qquad (6)$$

where C = material specific Curie constant, T = absolute temperature and theta ($\theta$) = Curie temperature. Curie temperature is the temperature at which certain materials loose their permanent magnetic properties, to be replaced by induced magnetism. In other words, it is the temperature above which a ferromagnetic material becomes paramagnetic. The Curie temperature is the temperature at which certain magnetic materials exhibit sharp change in their magnetic properties. The ferromagnetism has large magnetic susceptibilities, and saturation magnetization are obtained via external magnetic fields lower than that of paramagnetism. Frenkel and Dorfman [51] predicted the nature of ferromagnetism particle in the state of uniform magnetization at any applied field. By addition of external magnetic field, the domain leads to align along the field direction. The net magnetization does not become zero on the removal of magnetic field, which is useful for the fabrication of permanent magnets [52].

### 2.2.4. Antiferromagnetism

Antiferromagnetism establishes under the application of weak magnetic susceptibilities. Here, the atoms can be divided into sub-lattices in which magnetic dipole moments are aligned antiparallel, and form a low magnetic susceptibility. The temperature above which antiferromagnetism exits is called the Neel temperature denoted by $T_N$
materials have small positive susceptibility when compare with the paramagnetic materials.

### 2.2.5. Ferrimagnetism

The behaviour of ferrimagnetism is similar to that of ferromagnetism. The net magnitude of the ferrimagnetism and ferromagnetism are same during the arrangement of the magnetic dipole moment is very different. Further, it is divided into two sublattices and termed as a subset of antiferromagnetism. Each sublattice shows ferromagnetism nature, and only difference is that the magnetic dipole moments for the sublattices result in a net magnetization for ferrimagnetism.

The major examples of crystalline ferrimagnetism are double oxides of iron, as in MO, where M is the divalent material, and having a spinel-spinel structure (Sharma, 2017). The typical example is $Fe_3O_4$ (magnetite). With increase in temperature the movement of the molecules spins is disturbed through the thermal energy and result is decrease in magnetization. Furthermore, at Curie temperature this molecular arrangement/motion becomes completely disturbed, and magnetisation dissolves.

## 2.2.6. Superparamagnetism

This is a special type of magnetism. There are single domain particles exhibiting the nature similar to ferromagnetism below Curie temperature due to high magnetic susceptibilities. They are saturated in moderate magnetic field and show properties such as coercivity and remanence. They exhibit the characteristics similar to an ordinary paramagentism with no hysteresis beyond the Cuire temperature. The nature of superparamagnetism was proposed by Neel [53] who explained the thermal fluctuations in single-domain ferromagnetic particles.

The magnetic anisotropy energy of a particle is proportional to its volume i.e., $V$, the magnetic anisotropy energy is given by equation (7)

$$E_A = KV \sin^2 \theta \qquad (7)$$

where, $K$ is the anisotropy energy constant and $\theta$ the angle between the magnetization vector and easy axes of nanoparticles [54]. The comparison of different types of magnetism is reported in the literature (Palagummi and Yuan, 2016; Shatruk and Clark, 2023; Hedayatnasab et al., 2017).

The magnetic anisotropy energy moves towards its thermal energy when the volume of single-domain particles is very small. The magnetization flips between easy axes by an anisotropy barrier similar to classical paramagnetism system [55]. Since, it is associated with magnetic moment that of a single atom, the term superparamagnetism is applicable.

## 2.3. Coercivity

The coercivity, is a measure of the ability of the ferromagnetism to withstand under magnetic field without becoming demagnetized. It is similar to electric coercivity which measures the ability of a ferroelectric material to withstand against electric field without becoming depolarized.

The unit of coercivity is usually represented by oersted or ampere/meter and denoted by $H_C$. It can be measured using the instruments such as B-H analyzer or magnetometer. The magnetization of ferromagnetism is non-linear, and basically depends on the magnetic susceptibility and applied magnetic field as reported in [50]. It shows the relationship between external magnetic field ($\mu_0 H$) and total magnetic field $B$ or magnetization. Initially, the material is unmagnetized ($B = \mu_0 H = 0$). The magnetization saturates beyond the condition $B = \mu_0 H_s$, it means all domains point in the same direction. The ferromagnetism will retain a net magnetization when applied field moves towards zero. In order to bring the magnetization back to zero, applied magnetic field switched to the opposite direction. The required magnetic field that changes the direction of magnetization is known as the *coercive field* and denoted by $H_c$. A ferromagnetism sustains respective magnetization as long as the magnetic field is lower than its coercive field. Under this situation, ferromagnetism will align with magnetic moment that is parallel to the magnetic field.

## 2.4. Coercivity and Remanence in Magnetic Particles

The theory of magnetic domain explains that the critical size of single domain is affected by various factors which include the value of the magnetic saturation, the strength of the crystal anisotropy, surface energy and particles shape [56]. Schematic representation of hysteresis loop or curve between magnetization and magnetic field is available in literature [54]. Insertion of the magnetic field causes the spins within a particle causing an alignment with the field. The higher value of the magnetization is achieved in this condition which is known as saturation magnetization ($M_s$). Further, with the decrease in magnitude of the magnetic field, spin of the particles attains the previous condition, and again gets aligned with the field leading to decrease in the total magnetization. The ferromagnetism possesses residual magnetic moments at zero field, and its corresponding value of magnetization is called remanent magnetization ($M_r$). The coercive field ($H_c$) is the magnitude of the magnetic field which must be applied in the negative direction in order to bring the magnetization back to zero. The nature of the hysteresis loop is important for magnetic recording applications which require moderate coercivity, high remanent magnetization, and basically a square hysteresis loop [54].

The response of ferromagnetism is recorded against an applied magnetic field to get the hysteresis loop, which is explained by two distinct parameters: *coercivity and remanence*. These terms are subjected to the 'thickness' of the curve. The *coercivity* depends on the size of the particles. As the particle size decreases, the *coercivity* reached to its maximum and then decreases, ultimately moves to zero. The experimental validation of the importance of coercivity on particle size is presented in [53]. Energetic involvement favours

Introduction

the formation of domain walls in case of large size particles. The formation of domain walls becomes energetically unfavourable at the critical particle diameter $(D_c)$, and particles are called to be in single domain. The corresponding changes in the magnetization cannot be sustained longer due to domain wall and subsequently require the coherent rotation of spins, providing the large coercivities. If the particle size continuously decreases to below the single domain, their spin rotation is influenced by thermal vibrations, and the particle shows superparamagnetism as described by [54].

## 2.4.1. Terminology Involved in Magnetic Measurement

In case of ferromagnetism, the magnetic field intensity inside the sample is represented by $B$ and given as

$$B = H + 4\pi M \text{(in CGS) [57]} \tag{8}$$

$$B = \mu_0 + (H + M) \text{(in SI) [57]} \tag{9}$$

where, $M$ is the magnetisation induced within the sample by magnetic field $H$. For vacuum $M = 0$ under this condition equation (9) becomes

$$B \equiv \mu_0 H \text{ (in SI)} \tag{10}$$

$\mu_0$ is permeability of the vacuum and equal to
$\mu_0 = 4\pi \times 10^{-7} \, m \, kg A^{-2} s^{-2}$.

> **Oersted:** It is the CGS unit of auxiliary magnetic field ($H$-field) and denoted by Oe or $H$.
>
> **Gauss:** It is the CGS unit of measurement of magnetic flux density ($B$) and denoted by $G$.
>
> **Tesla:** Tesla is the SI unit of measurement of magnetic flux density ($B$) and denoted by $T$.
>
> Weber is the SI unit of measurement auxiliary magnetic field ($H$-field) and denoted by $W_g$.

Usually, magnetic field strength is measured by the unit of $Oe$ as $\left(\frac{A}{m}\right)$, and whenever it is measured by magnetic flux density, it is denoted by either $G$ or $T$. Magneto-optical properties of the materials [58] also provide intrinsic characteristics of the materials and their response under the magnetic field.

The hysteresis curve for a ferromagnetic material is discussed in [54]. It shows the meaning of two terms $H_c$ and $H_{ci}$, *coercivity* defined as the magnetization $M$ with applied field $H$ and magnetic flux density or induction $B$ with applied field $H$, respectively. Both the curves show similar nature and characteristic, but there is one basic difference.

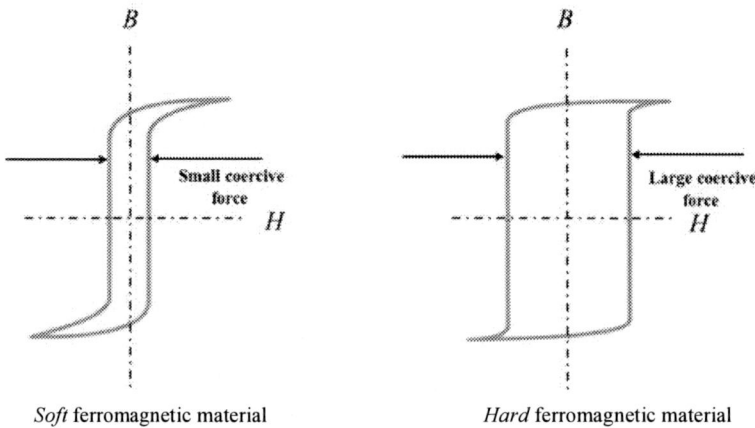

Soft ferromagnetic material          Hard ferromagnetic material

Ref: Ohring (1995); Palagummi and Yuan (2016).
Credit to: Ohring, M. Magnetic Properties of Materials. *Engineering Materials Science*, Academic Press, 711–746 (1995). https://doi.org/10.1016/b978-012524995-9/50038-6.
Palagummi, S.; Yuan, F. G. 8 - Magnetic levitation and its application for low frequency vibration energy harvesting, Editor(s): Fuh-Gwo Yuan, Structural Health Monitoring (SHM) in Aerospace Structures, Woodhead Publishing and Elsevier, The Netherlands. (2016). https://doi.org/10.1016/B978-0-08-100148-6.00008-1.

**Figure 2.1.** Hysteresis loop of soft and hard ferromagnetic material (Ohring, 1995; Palagummi and Yuan, 2016).

Beyond the saturation point ($Hsat$), the shape of curve $M$ changes to that of a straight line with zero slope. In second case, the slope of the curve $B$ signifies the constant magnetic susceptibility, and depends on the scale and units used to plot of $B$ Vs $H$. The $B$ Vs $H$ curve does not show the saturation by accessing a limiting value as in the case of the $M$ Vs $H$ curve. $M \equiv 0$ at $H = 0$ for the sample which is initially unmagnetized. $M$ and $B$ increases with H. The sample retains less magnetization when the applied field

H becomes zero, because of the presence of domains still aligned in the same direction as that of the applied field. The corresponding value at $H = 0$ is known as the *remnant magnetization* which is denoted by $M_r$ corresponding induction is knowing as *remnant induction* which is denoted by $B_r$. M and B to be zero, a reverse field is required which is known as the *coercive force* or *coercivity*. Coercive forces variation for *hard* and *soft magnetic* materials are compared for their small and large area of the hysteresis loop as shown in Figure 2.1.

## 2.5. Magnetic Nanoparticles

Nanomaterials are special class of particles with size range of nanometer which are synthesised from organic or inorganic materials [59]. Synthesis carried out by the addition of metal-organic precursors leading to formation of ultrafine and uniform nanoparticles [39, 60, 61, 62, 63, 64]. Magnetic nanoparticles have special collective and storage properties. Currently, these nanoparticles are one of the major sources for the preparation of magnetic fluids, data storage, targeted drug delivery, medical diagnosis and therapy, and several biotechnical applications such as bio separation and detection of biological species such as cell, protein, enzyme, bacteria, virus, etc. Concentration of the precursor and the reaction conditions have direct effects on the size of magnetic nanoparticles [65]. The controlled synthesis of magnetic nanoparticles under optimized conditions leads to the formation of $Fe_3O_4$, $\gamma$-$Fe_2O_3$ and $\alpha$-$Fe_2O_3$. $\gamma$-$Fe_2O_3$ is a metastable phase of $Fe_2O_3$, and directly transforms to $\alpha$-$Fe_2O_3$ or $\beta$-$Fe_2O_3$ or $\epsilon$-$Fe_2O_3$ [66, 67, FR 5].

Sukurai et al. [8] reported the transformation of four phases of $Fe_2O_3$ in the series of $\gamma$-$Fe_2O_3$ to $\epsilon$-$Fe_2O$ which transform to $\alpha$-$Fe_2O_3$. These transformations basically provide changes in the morphology or structure of the material which control the properties such as *saturation magnetization* $(M_S)$ and *coercivity* $(H_c)$ of these oxides. Magnetic nanoparticles for biomedical applications require quality particles with improved crystallinity and magnetic properties. Therefore, it is necessary to overcome the polydispersivity and heterogeneity for specific biomedical applications. The magnetic nanoparticles modified with ligands or polymers are found highly suitable as drug carrying vehicles. It has been observed that superparamagnetic particles easily accomplish these goals because they retard the agglomeration [68].

## 2.6. Magnetic Metal Oxide Nanoparticles (MMONPs)

MMONPs are popular as a wide range of smart and functional materials [69, 70, 71]. These materials consist of two different structures namely, mixed-valence cationic states and anionic deficiencies. The chemical, electrical and magnetic properties of oxide materials can be integrated [72] in order to tune the nanomaterials for desired applications (Hedaytnasab et al., 2017). The materials are placed in different metal groups with respective structures and properties. These nanoparticles usually consist of metal oxides e.g., iron oxides, nickel oxides and ferrites. Ferrites mostly consist of iron oxides and at least one metal in their chemical structure with specific magnetic and dielectric properties. They are generally expressed by $M$ ($Fe_xO_y$), where $M$ shows any metal cation e.g., manganese, magnesium, copper, nickel, cobalt, etc. and x and y values are variables [73, 74]. Due to special properties of ferrites, they are widely used for variety of applications such as biomedical imaging - magnetic resonance imaging, sensors, catalysts, and optical devices. Efficacy of these applications depends on dispersivity of nanoparticles because their size and shape greatly affect the speciality of applications.

It is observed that synthesised MMONPs have been partly unstable, and problems such as aggregation, hydrolysis and non-uniform shape and size affect their usage in various applications. Synthesis under vacuum or in non-aqueous medium is helpful to overcome these difficulties [75]. Due to their biological affinity and magnetic properties, $Fe_3O_4$, $\gamma$-$Fe_2O_3$ and ferrites with high oxidizing stability are recommended for biomedical applications. $Fe_3O_4$ particles show superior magnetic properties than that of other iron oxides. They contain $Fe^{2+}$ and $Fe^{3+}$ valence states of the inverse spinel-spinel structure, therefore $Fe_3O_4$ is more favourable substance for biomedical applications [74, 76].

## 2.7. Magnetic Alloy Nanoparticles (MANPs)

Metallic magnetic nanoparticles attracted a great research interest for their synthesis with monodispersity and narrow size distribution [77]. Three important methods involving chemical, physical and biological routes have been reported for synthesis of magnetic alloy nanoparticles [77, 78, 79].

Iron (Fe), nickel (Ni) and cobalt (Co) are among the most important magnetic alloy-based nanoparticles [80, 81]. The shape, size and compositions can be more easily controlled for these nanoparticles than that of metal oxide nanoparticles [82, 83]. The drawbacks of these nanoparticles are low chemical stability and biocompatibility, higher activity to oxidation, and ability to partially or completely magnetize due to hydrophobicity at room temperature [84]. Addition of other metals to produce magnetic alloy nanoparticles increases their resistance towards oxidation with improved magnetic properties as in Ni-Cr and Fe-Co-Au nanoparticles.

Wu et al. [85] suggested that carbon should be selected as the minimum preserver to produce Fe-based alloy nanoparticles. It improves mechanical and electrical properties. Diversified methods such as microemulsion combined with chemical reaction, thermal decomposition in organic or aqueous phase have been reported in the literature. Fe–Co coupled magnetic alloy has been frequently used due to magnetic properties such as high saturation magnetisation reported as 245 emu/g with negligible coercivity and high permeability [86]. These magnetic properties are of great importance for diversified applications such as high-density magnetic data storage, drug screening and delivery, but generally suffer from the ease of synthesis of these particles at nanosize range [87, 88, 89].

# Chapter 3

# Synthesis of Magnetic Nanoparticles

Magnetic nanoparticles have been synthesised through different metals, compositions and phases. They include iron oxide compounds, $Fe_3O_4$ and $\gamma$-$Fe_3O_4$, pure metal mainly Fe and Co, alloys likely $CoPt_3$ and $FePt_3$ as well as spinel shaped ferromagnets like $MgFe_2O_4$, $MnFe_2O_4$ and $CoFe_2O_4$. Synthesis of the magnetic nanomaterial has been achieved a remarkable attention by the researchers since last few decades. Enough literature is available explaining the efficient synthesis routes to produce controlled-shaped, highly stable and monodispersed magnetic nanoparticles.

Table 5.1 presents the overview of the preparation methods and their characteristics for the reported magnetic nanoparticles. The methods synthesis of the magnetic nanoparticles can be classified as under:

- Co-precipitation
- Thermal decomposition
- Hydrothermal synthesis
- Micro-emulsion
- Sol-gel reaction method

## 3.1. Co-Precipitation

The major advantages of the co-precipitation method are as under:

- High homogeneity
- Low cost
- Better product purity
- Less requirement of organic solvents

The magnetic nanoparticles can be prepared from aqueous salt solution by addition of a base in an inert atmosphere at room temperature [90]. The reaction involved in the co-precipitation method [91] is as mentioned in eq.

(11). The reaction pathway for synthesis of magnetite nanoparticles by co-precipitation is illustrated by Ahn et al. [92].

$$Fe^{2+} + 2Fe^{3+} + 8OH^- \rightarrow Fe(OH)_2 + 2Fe(OH)_3 \rightarrow Fe_3O_4 + H_2O$$

(11)

In the mid-early of 19$^{th}$ century, Bee [93] reported the synthesis of aqueous magnetic nanoparticles in acidic and alkaline phase through the co-precipitation method for synthesis of iron oxide magnetic particles. In this method, aqueous solution of iron oxide $Fe_3O_4$ or $\gamma$- $Fe_3O_4$ (salt solutions $Fe^{2+}/Fe^{3+}$) were mixed in their stoichiometric ratio with the addition of a base material under inert atmosphere at room temperature.

The tuning in the particle's properties such as size, shape and composition are generally based on following factors:

- Selected salts
- $Fe^{2+}/Fe^{3+}$ ratio
- Temperature
- pH of solution
- Ionic strength of the selected media

The size of particles decreases with the pH value and ionic strength. The synthesis conditions are established in this method, and the quality of particles is completely reproducible.

The stability of magnetic nanoparticles is less under ambient conditions. They are easily oxidized to maghemite and dissolved in acidic medium. Since the nature of maghemite is similar to ferrimagnet, they are less prone to get oxidized. This transformation is accomplished by dispersing the particles in acidic medium with the addition of iron (III) nitrate. The converted maghemite particles are chemically stable in both alkaline and acidic mediums. It was observed that the preparation of $Fe_3O_4$ nanoparticles using coprecipitation method under vacuum drying at 70°C is required for maintaining saturation magnetization [15]. Petcharoen and Sirivat [94] synthesized $Fe_3O_4$ nanoparticles with the addition of ammonium hydroxide as the precipitating agent during the coprecipitation method. The size and shape of $Fe_3O_4$

nanoparticles were basically controlled through changing the reaction temperature and surface properties. Peternele et al. [95] synthesized magnetite and maghemite nanoparticles using chemical coprecipitation of iron salts with addition of two alkaline chemicals namely, NaOH and NH$_4$OH as precipitating agents. The TEM analysis showed that the particles formed through a precipitating agent NH$_4$OH were more stable than that those achieved through NaOH. Roth et al. [96] studied various parameters affecting the coprecipitation and achieved the particles of size range from 3 nm to 17 nm, with saturation magnetization from 26Am$^2$ kg$^{-1}$ to 89 Am$^2$ kg$^{-1}$ by changing the iron salt concentrations, reaction temperature, and the ratios of hydroxide ions to iron oxide and $Fe^{2+}/Fe^{3+}$. Iida et al. [97] synthesised Fe$_3$O$_4$ nanoparticles using hydrolysis method in which aqueous solution consists of ferrous and ferric salts at different ratio with 1,6-hexanediamine as a base material. Their results confirmed that the formation of large hydroxide particles of Fe$_3$O$_4$ was accelerated and resulted in the size enlargement with an increase in the ratio of ferrous to ferric ions. The magnetic properties of nanoparticles can be controlled by changing the molar ratios of both the ions. The amorphous hydrated oxyhydroxide can be transformed to Fe$_2$O$_3$ at the temperature below 60°C. The formation of Fe$_3$O$_4$ initiates at temperature greater than 80°C. The size of particles decreases as the rate of mixing increases. Purging or bubbling of inert gas such as nitrogen through the solution prevents the critical oxidation of magnetic nanoparticles and also decreases the particle size.

Ahn et al. [92] presented the complete reaction pathways for coprecipitation of Fe$_3$O$_4$. The process starts with two nucleation steps as the reactant streams start to get mixed. The steps involved are "akaganeite → goethite → (hematite → maghemite) → magnetite" and "ferrous hydroxide → lepidocrocite → (maghemite) → magnetite" [3]. Generally, the magnetite formation steps coexist in co-precipitation method but in other processes the transformation of goethite to Fe$_3$O$_4$ is crucial leading to formation of arrow-shaped nanoparticles. Baumgartner et al. [98] studied the nucleation and particle growth mechanism of Fe$_3$O$_4$ nanoparticles in alkaline aqueous solutions. This mechanism is somewhat complicated and interrelated with co-precipitation method. These steps affect the particle size distribution, and responsible for low crystallinity of iron oxide nanoparticles synthesized through this method. Synthesis route to generate magnetic nanoparticles by co-precipitation method is explained in [93]. A significant advance for the preparation of monodispersed magnetic nanoparticles of various sizes and shapes have been made by the addition of organic additives as stabilizers or reducing agents. Chain-like cluster-precipitates of magnetic nanoparticles form on the addition of stabilizer or reducing agent.

The importance of various organic anions such as carboxylate and hydroxy carboxylate ions, during the formation of iron oxides is also an interesting topic of research. Recently, the oleic acid is found as the best candidate for the stabilization of $Fe_3O_4$ [99].

Qiu [100] studied the importance of ionic strength of the reacting solution for the formation of magnetic nanoparticles. The particles produced with the addition of 1M NaCl in aqueous solution were of smaller mean size than those formed without its addition. Even the particles formed with higher ionic strength showed comparatively lower saturation magnetisation. This lower magnetisation was referred to the consequences of decrease in size of the particles produced in higher ionic strength medium.

The major drawback of co-precipitation method is the agglomeration of particles due to extremely small particle size, resulting to high surface area and surface energy. Wu et al. [46] reported that maintaining the pH of solution during synthesis and purification stages is of very much importance for the formation of uniform and monodispersed nanoparticles.

## 3.2. Thermal Decomposition

Monodispersed nanoparticles having small size and shape can be synthesized through thermal decomposition of organometallic compounds in high-boiling organic solvents which consist of stabilizing agents [101, 102, 103]. The organometallic precursors consist of metal acetylacetonates, carbonyls, fatty acids, oleic acid and hexadecylamine. The proportions of basic materials containing the organometallic compounds, surfactant and solvent are basic parameters to control the size and shape of magnetic particles [104]. Other physical parameters such as reaction temperature and time also affect the size and shape of particles. Thermal decomposition of organometallic precursors with metal initially formed metal nanoparticles but due to the presence of oxygen it forms high quality monodispersed metal oxides. Pankhurst et al. [105] pointed out the uses of magnetic nanoparticles in biomedicine applications. MRI is dependent on particles size and therefore the synthesis of particles using this method could be beneficial for these applications.

Chen et al. [106] formed Nickel (Ni) nanoparticles through the thermal decomposition of Nickel (II) acetylacetonate using alkylamines. The morphology of the particles was controlled by the reaction temperature, time, rate of heating and choice of solvent. They observed that choosing the optimum reaction temperature and solvent compositions can help controlling the crystallinity of the synthesized magnetic particles.

Jana et al. [107] stated that in case of general decomposition method the size and shape of magnetic nanoparticles depend on the pyrolysis of fatty acid salts present in non-aqueous solution. These properties can also be correlated with the reactivity and concentration of the precursors (Ruiz and Bou). The reactivity depends on the chain length and the concentration of the fatty acids. Generally, the rate of reaction is faster with shorter chain length. Alcohols and primary amines were found to accelerate the reaction rate at low temperature range.

Park et al. [108] worked on preparation of iron oxide nanoparticles using thermal decomposition method. They selected nontoxic and inexpensive iron (III) chloride and sodium oleate to produce an iron oxalate complex which then decomposed at temperature range from 240°C to 320°C by addition of different solvents. The particle size range was from 5 to 22 nm. The 'aging' was found useful step for the synthesis of iron oxide nanoparticles. The particles synthesised using this process were highly dispersible in various organic solvents such as hexane and toluene. Water dispersible nanoparticles are widely useful for biotechnological applications. Water dispersible $Fe_3O_4$ particles were prepared by using $FeCl_3 \cdot 6H_2O$ and 2- pyrrolidone as coordinating solvent, under reflux at 245°C [109]. The average size of particles was found to be controlled at 4, 10 and 60 nm when the reflux time was 1, 10 and 60 h, respectively. The shape of the particles changed from spherical to cubic structure by changing the reflux time. The one-pot synthesis of water-soluble magnetite nanoparticles was reported under same conditions by the addition of capping agent. These nanoparticles were highly suitable as a magnetic resonance imaging contrast agent. Metal oxide magnetic nanoparticles can also be prepared through thermal decomposition method. Thermal decomposition of metal carbonyl precursors in the presence of air is observed. The decomposition of precursors with a cationic metal in the absence of reducing agents is observed. The disadvantage of this method is that organic soluble nanoparticles have a limited use in the biological applications. Furthermore, the surface modification is required after synthesis, and in some cases this method requires high temperature to get the particles dissolved in nonpolar solvents [46].

## 3.3. Hydrothermal Synthesis

Hydrothermal synthesis method is also known as solvothermal method for preparation of ultrafine powders and magnetic nanoparticles [110, 111, 112, 113]. An aqueous solution is maintained at a pressure 2000 psi and temperature more than 200°C in this method. This method is found convenient for growing crystals of different materials. Wang et al. [113] worked on hydrothermal method, and generalised different types of nanocrystals by a liquid-solid-solution reaction. The hydrothermal reaction occurs in phases such as sodium linoleate (solid), ethanol-linoleic acid (liquid phase) and water-ethanol at various temperatures. The reaction is carried out at specific phase transfer and separation at the interfaces of liquid, solid and solution phases available during synthesis. Wang et al. [113] reported that the multicomponent mixture like ethylene glycol, sodium acetate, and polyethylene glycol, to direct synthesis of particles. Ethylene glycol have high boiling point and worked as reducing agent which is widely applicable for polyol process to form monodisperse metal or metal oxide nanoparticles. Sodium acetate works as electrostatic stabilizer to overcome particle agglomeration and polyethylene glycol is a surfactant agent that prevent particle agglomeration [FR6].

Wang et al. [114] used hydrothermal method to synthesised $Fe_3O_4$ nanoparticles. They achieved the saturation magnetisation of 85.8 emu $g^{-1}$ for their $Fe_3O_4$ particles of average size 40 nm. This is less than that of corresponding bulk magnetisation of $Fe_3O_4$ (92 emu $g^{-1}$). It was found that the structure of the $Fe_3O_4$ particles formed under specific hydrothermal conditions affects the saturation magnetisation. Nickel ferrite nanoparticles of size 6-170 nm synthesized using hydrothermal method incorporating electrostatic stabilization in ethylene glycol and NaAc have revealed that the processing conditions such as reaction time, initial concentration, amount of selected reagents and the type of acetates affect the particle size distribution [115]. A generalized step of hydrothermal method for synthesis of nanomaterial is illustrated in [111].

Xu and Teja [116] explained various types of parameters that affects particle size, its distribution, and morphology of α-$Fe_2O_3$ nanoparticles produced through continuous hydrothermal synthesis. They found that the aggregation of particles can be prevented by the addition of polyvinyl alcohol during synthesis. Further, the particle size distribution was found to be influenced by the processing temperature and residence time. Even though the hydrothermal method looks promising, the main disadvantage of this method is its reaction kinetics. Microwave heating is found to improve the reaction rate [117], and such a hybrid method is known as microwave-hydrothermal assisted method. As pointed out, the addition of microwaves accelerates the reaction kinetics by providing energy for particle synthesis.

Sreeja and Joy [118] synthesised $\gamma$-$Fe_2O_3$ nanoparticles using microwave-hydrothermal assisted method. They found this single step method as a simple and fast alternative route for the synthesis of magnetic nanoparticles. $\gamma$-$Fe_2O_3$ nanoparticles were synthesized at 150°C within a shorter period of time, i.e., 25 min as compare to convectional hydrothermal methods which usually take longer time.

Hydrothermal method is comparatively less preferable for the synthesis of magnetic nanoparticles. However, this method provides high-quality magnetic nanoparticles. Till now co-precipitation and thermal decomposition are among the preferable methods for synthesis of magnetic nanoparticles with high throughput. Major concern is the stability of the nanoparticles as it with other colloidal dispersions. The stability of nanoparticles is governed by the electrostatic repulsion which depends on type of stabilizer used. Repulsive electrostatic force due to positive charge on the particles have been found useful for stabilizing magnetic nanoparticles synthesized through co-precipitation [93]. Organic solvents containing fatty acids or surfactants have shown the remarkable stability for the nanoparticles prepared by thermal decomposition method [63].

## 3.4. Microemulsion

Microemulsion is also a competitive method in which water-in-oil emulsion facilitates the synthesis of uniform sized magnetic nanoparticles [119, 120, 121, 122]. Emulsion is an isentropic and thermodynamically stable system contains three components; water, oil and surfactant. Surfactants reduce the surface tension between aqueous and organic phases. Water-in-oil emulsions contain aqueous droplets dispersed in continuous organic phase. Here, the aqueous droplets constitute a discrete phase. On the other hand, oil-in-water emulsion contains discrete droplets of organic liquid dispersed in continuous aqueous phase. The dispersed entities sometimes precipitate in the form of micelles [123]. The precipitation reaction and aggregation process facilitate the synthesis of magnetic nanoparticles. The dispersed droplets serve as a tool to offer a room for the reaction to take place within them. Such type of microreactor helps controlling the size of the nanoparticles being synthesized within them. Solvents, such as acetone or ethanol are usually used from which the precipitates can be separated through filtration or centrifugation of the dispersions. Abou-Hassan et al. [124] reported the synthesis of magnetic and fluorescent $\gamma$-$Fe_2O_3$@$SiO_2$ nanoparticles using the microreactor. The microreactors can achieve chemical synthesis within their tiny volumes and improve heat and mass transfer. The microreactors can be designed in such a way so as to provide continuous, multistep, series of reactions without leaving the microreactor environment. The mixing, grafting and coating are performed inside the microreactor with better control of the heat and mass transfer. Mixing occurred by diffusion at the interface of the reactants. The synthesised particles have uniform size and shape which can be obtained on comparatively large scale in a time effective method. The influence of precursor concentrations and residence time on coating are also important to consider for effective synthesis of the particles. Magnetic nanoparticles can be further modified and functionalized to improve their dispersibility in biological fluids for specific applications.

Santra et al. [121] reported a technique for the synthesis of coated and uncoated silica nanoparticles uniformly distributed through water-in-oil emulsion. They combined three different non-ionic surfactants for the preparations of emulsions using $NH_4OH$ and $NaOH$ as base material. Water-in-oil emulsions were formed by mixing two identical reagents in which one of them was metal salt and the other was the base material. They found that the chemical structure or nature of the surfactant molecules affected the rate of surfactant adsorption onto the surface of nanoparticles.

The nanodroplets of water which contains reagents that prevent fast coalescence and aggregation phenomena have been reported for the synthesis of magnetic nanoparticles [121]. Vidal-Vidal et al. [37] synthesized monodispersed magnetic nanoparticles using microemulsion method. The spherical particles coated with a monolayer of oleic acid; a narrow size distribution of 3.5 ± 0.6 nm, were not completely crystallized have shown large saturation magnetisation. It was also observed that oleic acid worked as a precipitating and capping agent. Microemulsions mostly used to synthesized monodispersed nanoparticles with various geometries. Low yield and large amount of solvent usage are the major drawbacks of these techniques.

## 3.5. Sol-Gel Reaction Method

This is an ancient method in which hydrolysis and condensation of metal alkoxides. It involves the formation of dispersed metal oxide particles in sol, followed by drying or gelling to remove solvent. Sol is the stable dispersion of colloidal particles in the solvent. A gel contains three-dimensional continuous network entrapping a liquid phase. Colloidal gel is made from the agglomeration of colloidal particles. Basically, this method follows four steps:

- Hydrolysis
- Condensation
- Drying
- Thermal treatment

Generally, sol particles exhibit van der Waals forces or hydrogen bond. Gel structure may also form due to linking polymer chains. Precursors used for synthesis of iron oxide nanoparticles are iron alkoxides and iron salts such as chlorides, nitrates and acetates. These precursors undergo for polycondensation and hydrolysis reactions at room temperature which require heat treatment to achieve the final crystalline structure.

Cui et al. [6] reported the reactions between $FeCl_2$ and propylene oxide in ethanol solutions at their boiling point leading to sol formation and continuous drying. The phases of iron oxide can be formed by changing the drying conditions for sol. Lemine et al. [125] synthesized $Fe_3O_4$ nanoparticles with an average particle size of 8 nm using this method. The saturation magnetisation achieved up to 47 emu g$^{-1}$ at room temperature, and confirmed that these nanoparticles had great potential in the biomedical field. Qi et al. [126] synthesised $Fe_3O_4$ nanoparticles in the size range from 9 to 22 nm by a non-alkoxide as a precursor using sol-gel method. The shape and crystal structure of iron oxide nanoparticles depend on the choice of organic precursors. Woo et al. [127] synthesized $Fe_2O_3$ nanorods in the controlled phase at atmospheric conditions. Chemical components, $H_2O$/oleic-acid ratio could be used to control length and diameter of the nanorods, and their structure could be controlled by physical parameters such as temperature and hydrous state of the gels during crystallization.

The co-precipitation method has several advantages over the sol-gel method. The iron oxide nanoparticles can be firmly dispersed in the aqueous medium and in several polar solvents due to surface characteristics of the synthesized oxide particles which contain various hydrophilic ligands. This method formed iron oxide magnetic nanoparticles with higher crystallinity and saturation magnetisation due to high relative temperature. The comparatively high cost and release of the large amount of alcohols during heating process inviting special safety considerations are the major disadvantages of sol-gel method.

# Chapter 4

# Magnetic Properties of Magnetic Particles

The physiochemical characteristics of nanoparticles depend on their size. Particles having size range of micron or more can comparatively easily synthesized. The intensity of magnetic particles decreases in the absence of external magnetic field, and energetically beneficial as compared to homogenously magnetized particles [128]. The magnetization directions of all particles fall in the domain are statically oriented, which results in the formation of magnetic moments within particle in the absence of magnetic field.

As the size of the particles decreases, the relative proportion of wall energy to that of whole particle energy increases. Therefore, no magnetic domains are established below to critical particle size, and whole particles possess a natural magnetization in one direction. It was showed by Dutz et al. [129] that for spherical and cubic particles made of magnetite, the hypothetical range for the generation of single domain particle was nearly 80 nm [130].

The magnetization of small magnetic particles is basically related to their measurement time and Neel relaxation time. If the measurement time is so less than relaxation time; particles have enough time for relaxation. In the absence of external magnetic field, if the time used to measure the magnetization of the nanoparticles is sufficiently more than the Neel relaxation time, the average magnetism on the particles vanishes leading to the exhibition of the superparamagnetism. Superparamagnetism of particles have specific collective properties with no coercivity but shows strong hysteresis when subjected to highly intense alternating magnetic field or magnetic field strength.

A special category of magnetism formed when small superparamagnetic particles form clusters. These clusters show superparamagnetism with no coercivity under the absence of external magnetic field. Depending on the interparticle interaction, a collective magnetism may be exhibited, and the clusters show ferrimagnetism with a reasonable hysteresis in the presence of external magnetic field. It is well known that superferrimagnetism and ferrimagnetism of the particles are favourable properties for magnetic applications [129, 131]. It is evident that particle size imparts important role for the magnetic nature of nanoparticles. The size distribution is an another factor affecting the characteristics of the magnetic nanoparticles [84].

# Chapter 5

# Characterization of Particles

## 5.1. Techniques of Characterization

Particle size and particle size distribution largely influence the characteristic of any nanomaterials or particles. So that it is crucial to identify these parameters. Dynamic light scattering (DLS) is a very useful tool for particle size characterization. DLS uses the Brownian motion to allow information about the hydrodynamic radius, size distribution (polydispersity index - PDI) and colloidal stability of nanoparticles in solution. PDI values from 0.1-0.25 indicate the narrow size distribution, and PDI values above 0.5 indicate broad size distribution. The size distribution determined from the DLS is very easy method and important technique to know the effect of surface modification or encapsulation of the nanoparticles. It is also helpful to predict the stability of the dispersion. The zeta potential is an important parameter to interpret the stability of colloidal dispersions. However, DLS technique make use of light scattering due to the presence of nanoparticles in its path. Hence, the proper measurement technique, refractive index, concentration of the sample used for the measurement also play an important role during the DLS analysis. It is also advisable to use suitable imaging technique in addition to the DLS technique in order to correctly interpret the results of particle size distribution. Moreover, DLS provides the size distribution of the particles present in the samples from which the mean particle size is known. On the other hand, electron microscopy can provide the information of size and shape of individual particles depending on the compatibility of the instruments. Transmission electron microscopy (TEM) provides the information about size, shape and thickness of the individual particles. TEM information can easily be used to quantify the effect of polymeric shell on the agglomeration tendency of the magnetic nanoparticle case of core-shell structure. This is very important due to the consequences of rupturing of organic nanostructure which may occur during drying process which may otherwise lead to wrong interpretation. Due to this reason often DLS and TEM are reported in combination. Field emission scanning electron microscopy can also provide the information about the size and morphology of the sample.

Powder X-ray diffraction (P-XRD) analysis is generally used to identify the crystal structure and phase of magnetic core. This analysis gives information related to crystallinity of the particles as well as mean value of the particle diameter [132]. Vibrating sample magnetometer (VSM) is used to measure the magnetic properties of nanoparticles. Presence of diamagnetic material can also be confirmed from the magnetic properties. In addition, this method can also confirm whether the analysed particles are superparamagnetic or not.

## 5.2. Magnetism and Microfluidics

Presently, magnetic forces can be combined with microfluidics in an incredible variety of ways. Generally, magnetic forces are applicable to steer magnetic objects such as magnetic particles [133] magnetically labelled cells [134], and plugs of ferrofluid inside a microchannel. Besides these, magnetic forces are frequently utilized to manipulate non-magnetic particles (diamagnetic objects) [135]. Sophisticated microfabricated electromagnets are used to apply external magnetic force through the microchannels. Magnetic nanoparticles present in the stream flowing in the microchannel experience the magnetic force, and thus guided steering, trapping or separation and sorting of magnetic particles can be achieved in a very controlled manner due to the influence of external magnetic field. Bio-assay have been developed on the surface of the magnetic particles and captured within a microchannel. Applied external magnetic forces can control the magnetic object inside the microchannel without the direct contact with the fluid. Even biomolecules can be isolated or transferred from the sample with precise control by selectively adhering them to small magnetic nanoparticles, and can be further removed using external magnetic field [49].

## 5.3. Characterization of Magnetic Nanoparticles

The XRD analysis is a basic characterization technique for identifying the crystal structure as well as classifications of magnetic nanoparticles. It is important to remember that heat treatment is an essential step in synthesis of crystalline structure of particles.

## 5.4. TEM and SEM Analysis

TEM micrograph shows the morphology and size distribution of nanoparticles. It is used to get the confirmation of the shape and size of the synthesised nanoparticles, and to check the match with the data obtained from XRD analysis. SEM analysis is used to identify the morphology of the synthesised nanomaterials. The images obtained using selected area electron diffraction (SAED) can be used to identify the aggregation. Magnetic hysteresis loop or $M$-$H$ curve which is based on magnetic nanoparticles is explained in literature [15]. The delay in the response of magnetic nanoparticles known as magnetic hysteresis which is related to their magnetization. Through magnetization, material initially becomes magnetized and later on demagnetized. $α$-$Fe_2O_3$ shows weak ferromagnetism at room temperature, and its saturation magnetization value is < 1 emu/g. On the other hand, $γ$-$Fe_2O_3$ and $Fe_3O_4$ shows ferrimagnetism at room temperature, and saturation magnetization values reach up to 92 emu/g which is in line with the magnetization of bulk iron oxide particles [15]. Moreover, the residual magnetization and coercivity can also be found from the $M$–$H$ curves [136]. There are two key parameters that dominates the magnetic properties of iron oxide magnetic nanoparticles which are finite-size and surface. The finite-size affects the surface-to-volume ratio and crystal structure. During large scale production of magnetic nanoparticles, presence of a multidomain structure and regions of uniform magnetic field are separated by domain wall. At nanoscale iron oxide magnetic nanoparticles generally shows single-domain alignment. Since single-domain particle shows uniform magnetization, all spins aligns in the same direction. Therefore, the larger value of coercivity is observed in small size of iron oxide nanoparticles. A special property of iron oxide magnetic nanoparticles is the superparamagnetism. Iron oxide magnetic nanoparticles shows superparamagnetism as the $M$-$H$ loop does not represent hysteresis because forward and backward magnetization curves mitigate with each other. Magnetic nanoparticles have zero magnetization in the absence of external field, and no remnant magnetism is appeared on removal of applied field. [137].

## 5.5. Size Dependent Characteristics of Iron Oxide Nanoparticles

The size of nanoparticles is important to explore size-dependent magnetic properties. Park et al. [108] synthesized monodispersed iron oxide nanoparticles with size range 6-13 nm, and saturation magnetisation was found to increase with grain size of magnetic nanoparticles. Xuan et al. [138] prepared $Fe_3O_4$ nanoparticles with an average size of 280 nm using hydrothermal method. The initial particle size was controlled from 5.9 to 21.5 nm by changing weight ratios of feedstocks. Further, Demortière et al. [139] synthesised monodispersed magnetic nanoparticles using an iron-oleate precursor, and found the increase in average diameter with the increase of blocking temperature. Magnetic properties of iron oxide nanoparticles also depend on their compositions [140]. Yun et al. [141] reported that properties of magnetite-rich nanocrystals did not depend on their size but that for maghemite nanocrystals was completely particle size dependent. Guardia et al. [142] found comparatively higher specific absorption rate for nanoparticle of size 35 nm as compared to that for19 nm and 24 nm nanoparticles.

## 5.6. Shape-Dependent Characteristics Iron Oxide Nanoparticles

Besides spherical magnetic nanoparticles, nanorods and nanotubes are also very common nowadays for magnetic nanoparticles. Liu et al. [143] prepared $α-Fe_2O_3$ nanorods and nanotubes through hydrothermal method. The nanorods showed a transition at 166 K from the canted antiferromagnetism phase while nanotubes showed a 3D magnetic properties above 300 K [143]. The dynamic and kinetic parameters are responsible for the nucleation and growth by changing the reaction conditions affect the shape of nanoparticles [144]. The cubic $α-Fe_2O_3$ showed superparamagnetism at room temperature. The orthorhombic $α-Fe_2O_3$ nanoparticles showed ferromagnetism and low-temperature phase transition as reported by Wu et al. [144].

Song et al. [145] produced quasi-cubical shaped nanoparticle using shorter surface exchange time, and exhibited that stronger saturation magnetization helped to improve contrast for magnetic resonance imaging as well as enhanced conversion efficiency under magnetic induction. The crystal-

## 5.7. Effects of Agglomeration of Magnetic Nanoparticles

Nanoparticles have higher surface energy due to their large surface to volume ratio. Therefore, they are prone to occupy stable position close to each other in order to reduce surface energy [146]. Magnetic nanoparticles have high potential to agglomerate due to existence of magnetic dipole-dipole attraction [147]. Phenrat et al. [148] showed the transformation of magnetic iron nanoparticles into agglomeration. Henry et al. [149] explained the agglomeration of nanoparticles as it consists of *collision* and *adhesion* steps. In *collision*, particles travel within the fluid and collide with each other where as in *adhesion* step, particle-particle interactions play an important role. During *collision, the* colloidal particles may combine together (agglomeration) or remain separated (nonagglomeration).

Agglomeration can also be explained by different approaches. In the presence of the external magnetic field, the magnetic domain aligns in the direction of applied field, and reaching towards saturation magnetization. On removal of magnetic field, the domain will turn back to their initially oriented state which shows no macroscale magnetism.

## 5.8. Colloidal Stability of Magnetic Nanoparticles

Magnetic nanoparticles can sharply loose their colloidal stability and show the tendency to agglomerate [150]. Several approaches have been suggested to enhance the colloidal stability of magnetic nanoparticles as shown in [149-150]. One of the convenient and faster method to improve the dispersibility of magnetic nanoparticles is using mechanical sonication with proper selection of sonicator type, sonication duration and output power. The vibrational waves initiated by sonication improve the dispersibility of the nanoparticles. These waves create acoustic cavitations that initiate the formation and collapse of small bubbles in the solution. A localized hot spot formed naerby the suspended nanoparticles leads to increase the temperature and pressure. This ultimately results in disintrigation of large aggomerated particles into smaller clusters as reported by Dickson et al. [151]. Wu et al. [152] successfully decreased the aggomeration of $TiO_2$ nanoparticles using probe sonicator. The probe sonication was used at 20 W for 5 min to produces $TiO_2$ nanoparticles in range of 600 nm. Furthermore, with increase in sonication time the particle size decreased significantly, for instance, sonication for 30 min period decreased the particle size to 200 nm. It was also observed that the polydispersity index of $TiO_2$ nanoparticles was significantly reduced from 0.3 to 0.15 during 30 min sonication period [150], showing much improvement in suspension monodispersity.

### 5.8.1. Surface Coatings of Nanoparticles

Surface modifications or coatings remarkably improve the colloidal stability of nanoparticles. Several chemicals have been used to offer steric, electrostatic or electrosteric stabilization. Natural polyelectrolytes such as humic acid and strach have potential for stabilization of nanoparticles [151, 153, 154]. The phynyl compounds and carbon chains provide steric repulsion between magnetic nanoparticles [155].

The functional groups form bonding onto the surface of magnetic nanoparticles [155, 156, 157]. The interaction of the functional groups on the surface improves the surface properties magnetic nanoparticles. Surface coating of chitosan and silica particles have been reported to show the improvement in adsorption and selectivity towards metal ions [158, 159, 160, 161]. Many researchers have worked on functional polymers to synthesize magnetic nanoparticles. Nanoparticle modifiers such as polyacrylic acid (PVA) reported by [162, 163, 164, 165], and carboxymethyl cellulose (CMC) reported by [158, 159, 166, 167] have been found to improve the stability of magnetic nanoparticles.

However, depending on the quality of the coating, the dispersibility of the magnetic nanoparticles is found to be affected predominantly due to the characteristics of the coatings. The modifiers firmly adsorbed on the surface of particles with some complexion due to the concentration gradient. However, these modifiers may be reversible in atmospheric conditions in diluted form with higher pH and ionic strength [168]. Liu et al. [169] synthesized magnetic nanoparticles at low pH and high ionic strength. They observed that the stability of synthesised particles changed as the coating material desorbed.

**Table 5.1.** Overview of reported magnetic nanoparticles preparation with their specific properties

| Sr. No. | Reagents | Method of preparation | Nature of particles | Temperature | Size and shape | Area of applications | Major outcomes | Ref. |
|---|---|---|---|---|---|---|---|---|
| 1. | $(NO_3)_3 \cdot 9H_2O$, Polyvinyl alcohol | Hydrothermal | PVA coated iron oxide nanoparticles | 300°C | 15.6-22 nm rhombic | Catalyst, inorganic pigments, magnetic recording media | The presence of PVA during synthesis hinders particle aggregation, and thereby results in uniform particles and narrow particle size distributions | [116] |
| 2. | Tetraethoxysilane (TEOS), Titanium tetraisopropoxide (TTIP) in isopropyl alcohol (IPA) containing octadecylamine (ODA), Ammonia ($NH_3$) | Sol–gel | Silica–titania hybrid functional nanoparticles | Room Temp. | 40 nm | High sensitivity towards humidity sensor | Control synthesis of hybrid nanoparticle through microfluidic approach | [173] |
| 3. | $FeCl_2 \cdot 4H_2O$ in HCl, $FeCl_3$ in HCl, NaOH | Co-precipitation | $Zn_{0.37}Fe_{2.63}O_4$, $Zn_{0.48}Fe_{2.52}O_4$ | | 1.6 nm, 1.3 nm | Preparation of Zn doped $Fe_3O_4$ nanoparticles, Zn-substituted $Fe_3O_4$ nanoparticles with linear increase of Zn content | Zinc-doped $Fe_3O_4$, $Zn_xFe_{3-x}O_4$ ($0 \leq x \leq 0.48$), nanoparticles using microfluidic technique | [174] |

**Table 5.1. (Continued)**

| Sr. No | Reagents | Method of preparation | Nature of particles | Temperature | Size and shape | Area of applications | Major outcomes | Ref. |
|---|---|---|---|---|---|---|---|---|
| 4. | Iron (III) chloride hexahydrate (FeCl$_3$.6H$_2$O) Iron (II) chloride tetrahydrate (FeCl$_2$.4H$_2$O), Ammonium hydroxide (NH$_4$OH, 25%), Hydrofluoric acid (HF, 48 %), Nitric acid (HNO$_3$, 69%) | Co-precipitation | Fe$_3$O$_4$@ MIL-100 (Fe) | 80°C | 10–20 nm crystalline | Cancer treatment | Fe$_3$O$_4$ nanoparticles into porous MIL-100 structure for better loading efficiencies and with controlled release capabilities | [175] |
| 5. | Ferric chloride, Ferric nitrate, Cationic surfactant cetyltrimethylammonium bromide (CTAB), Anionic surfactant sodium dodecylsulfate (SDS), Tetramethylammonium hydroxide (TMAOH, 25% aqueous) | Hydrothermal | Nano crystalline α-Fe$_2$O$_3$ | 80°C | 17–20 nm, rectangular shaped particles | Magnetic devices, anticorrosive agent | The magnetization showed the branching between the field cooled and zero field cooled magnetization up to 340 K | [112] |

| Sr. No. | Reagents | Method of preparation | Nature of particles | Temperature | Size and shape | Area of applications | Major outcomes | Ref. |
|---|---|---|---|---|---|---|---|---|
| 6. | $(NH_4)_2Fe(SO_4)_2$, $FeCl_3$, NaOH | Co-precipitation | $Fe_3O_4$ magnetic nanoparticles | 80°C | 10 nm Cubic single phase nano size | Drug delivery systems (DDS), medical applications | The particles is consider to be superparamagnet and has no hysteresis loop | [176] |
| 7. | Ammonium hydroxide Iron (II) chloride Iron (III) chloride | Co-precipitation | $\gamma$-$Fe_2O_3$ particles | 90°C | 8 nm spherical and poly disperse | Medical imaging, biomedical | Origin of the dispersion bio-distribution of the particles, and small magnetic particles are at the present time extensively studied by specialists in medical imaging | [93] |
| 8. | Ferric chloride ($FeCl_3 \cdot 6H_2O$), Ferrous sulfate ($FeSO_4 \cdot 7H_2O$), Aqueous ammonia hydroxide ($NH_4OH$), Hydrazine hydrate ($N_2H_4 \cdot H_2O$), Anhydrous ethanol ($C_2H_5OH$), Sodium oleate ($C_{17}H_{33}COONa$), Polyethylene glycol-4000 (PEG-4000) | Continuous microwave irradiation | $Fe_3O_4$ nanoparticle, oleate-coated $Fe_3O_4$ nanoparticles for preparation of magnetic fluid | 80°C | 10 nm quasi-spherical | Cancer therapy, magnetic fluid separation | The nanoparticles synthesized using $NH_4OH$ as precipitator with hydrazine demonstrated good dispersion, while the prepared MFs revealed excellent stability and high susceptibility | [177] |

**Table 5.1.** (Continued)

| Sr. No. | Reagents | Method of preparation | Nature of particles | Temperature | Size and shape | Area of applications | Major outcomes | Ref. |
|---|---|---|---|---|---|---|---|---|
| 9. | Ferrous sulfate ($FeSO_4 \cdot 7H_2O$), Ferric sulfate ($Fe_2(SO_4)_3 \cdot nH_2O$), Ferrous chloride ($FeCl_2 \cdot 4H_2O$), Ferric chloride ($FeCl_3 \cdot 6H_2O$), 1,6-hexanediamine ($H_2N(CH_2)_6NH_2$) | Hydrolysis of solutions containing various iron salts | $Fe_3O_4$ nanoparticles | NA | ~9 to ~37 nm | Biological and biomedical fields | Magnetic properties of $Fe_3O_4$ nanoparticles can be controlled by adjusting the molar ratio of ferrous to ferric ions as well as the particle diameter | [97] |
| 10. | $FeCl_3 \cdot 6H_2O$, $FeCl_2 \cdot 4H_2O$, NaOH, Sodium citrate ($C_2H_5ONa_3C_6H_5O_7 \cdot 2H_2O$), Oleic acid ($C_{17}H_{33}COOH$) | Co-precipitation | $Fe_3O_4$ magnetic nanoparticles | 40-80°C | 12-15 nm spherical shape | Biomedical applications | $Fe_3O_4$ MNPs modified by sodium citrate and oleic acid show excellent dispersion capability, which should be ascribed to the great reduction of high surface energy and dipolar attraction of the nanoparticles | [178] |
| 11. | Iron chlorides ($FeCl_3$ and $FeCl_3 \cdot 6H_2O$), Iron nitrate ($Fe(NO_3)_3 \cdot 9H_2O$), Sodium hydroxide (NaOH), 25 % ammonia solution, Potassium iodide (KI), Polyvinyl alcohol (PVA 72000) | Co-precipitation | Magnetite nanoparticles | Room temp. | $7.8 \pm 0.05$ nm and $6.3 \pm 0.2$ nm nanocubes and nanorods composite | Storage devices, rotary shaft sealing, position sensing | A high selectivity and atom economy percents were achieved indicating that the method is environmentally benign and green | [179] |

| Sr. No. | Reagents | Method of preparation | Nature of particles | Temperature | Size and shape | Area of applications | Major outcomes | Ref. |
|---|---|---|---|---|---|---|---|---|
| 12. | Ferric chloride (FeCl$_3$), Ferrous chloride (FeCl$_2$), Sodium hydroxide (NaOH) | Co-precipitation | Fe$_3$O$_4$ magnetic nanoparticles | 70°C | 5-20 nm spinel structure | Remove dye in the water by a simple magnetic separation process | The optimum adsorption occurred at initial concentration of procion dye 100 mg L$^{-1}$, pH solution was 6, dosage of Fe$_3$O$_4$ 0.8 g L$^{-1}$ and contact time 30 minutes adsorption capacity was 30.503 mg g$^{-1}$ | [180] |
| 13. | FeCl$_3$·6H$_2$O, NaH$_2$PO$_4$·2H$_2$O, Na$_2$SO$_4$ | Hydrothermal | Maghemite ($\alpha$-Fe$_2$O$_3$) | 220°C | capsule-like tubular short nanotubes | Biotechnology, biomedicine | The self-assembly strategy is an efficient way to create novel nanostructured systems | [136] |
| 14. | Ferric chloride hexahydrate (FeCl$_3$·6H$_2$O), Ferrous chloride tetrahydrate (FeCl$_2$·4H$_2$O), Sodium oleate (C$_{18}$H$_{33}$NaO$_2$), Sodium hydroxide (NaOH) hydrochloric acid (HCl) | Chemical solution | Mono-dispersed iron oxide nanoparticles | | 6 nm oval to sphere | Medical and physical | Super paramagnetism with a blocking temperature around 150 K, and almost immeasurable remanence and coercivity | [181] |

**Table 5.1.** (Continued)

| Sr. No. | Reagents | Method of preparation | Nature of particles | Temperature | Size and shape | Area of applications | Major outcomes | Ref. |
|---|---|---|---|---|---|---|---|---|
| 14a | Ferric chloride hexahydrate (FeCl$_3$·6H$_2$O), Ferrous chloride tetrahydrate (FeCl$_2$·4H$_2$O) | Co-precipitation | Magnetic nanoparticles | 80°C | 25-30 nm | Magnetic nanofluid | Magnetic nanoparticles were dispersed in the dispersion media and magnetic nanofluid was prepared | [182] |
| 15. | Ferrous chloride tetrahydrate, Ferric chloride anhydrous, Ammonium hydroxide, Oleic acid, Hexanoic acid, Ethanol | Co-precipitation | Magnetite nanoparticles | 80°C | 10–40 nm, spherical-shaped | Drug release, cancer therapy, hyperthermia | Magnetite nanoparticles showed the super-paramagnetic behaviour with high saturation magnetization | [94] |
| 16. | FeCl$_3$·6H$_2$O FeCl$_2$·4H$_2$O, NaOH | Co-precipitation | Fe$_3$O$_4$ magnetic nanoparticles | 80°C | 9.64 nm, cubic | Ferrofluid preparation | Super paramagnetic behaviour at room temperature, documented by the hysteresis loop recorded | [132] |

| Sr. No. | Reagents | Method of preparation | Nature of particles | Temperature | Size and shape | Area of applications | Major outcomes | Ref. |
|---|---|---|---|---|---|---|---|---|
| 17. | $FeCl_2 \cdot 4H_2O$, $FeCl_3 \cdot 6H_2O$, Aqueous NaOH, Sodium oleate, polyethylene glycol (PEG) | Co-precipitation | Biocompatible $Fe_3O_4$ nanoparticles | | 8–20 nm | Magnetic drug targeting, magnetic resonance imaging | $Fe_3O_4$ nanoparticles coated by sodium oleate had a better biocompatibility, better magnetic properties, easier washing, lower cost, and better dispersion than the magnetite nanoparticles coated by PEG | [183] |
| 18. | Nickel chloride hexahydrate ($NiCl_2 \cdot 6H_2O$), Iron chloride hexahydrate ($FeCl_3 \cdot 6H_2O$), Sodium acetate ($NaAc \cdot 3H_2O$), Ethylene glycol | Solvothermal | Nickel ferrite nanoparticles | Room temp. | 130-240 nm spheres | Magnetic resonance imaging, targeted drug delivery | Coactivity and magnetization values are directly influenced by the size of final products | [115] |

**Table 5.1.** (Continued)

| Sr. No. | Reagents | Method of preparation | Nature of particles | Temperature | Size and shape | Area of applications | Major outcomes | Ref. |
|---|---|---|---|---|---|---|---|---|
| 19. | Ferric chloride hexa-hydrate ($FeCl_3 \cdot 6H_2O$), Ferrous chloride tetrahydrate ($FeCl_2 \cdot 4H_2O$), Propylene glycol ($CH_3CH(OH)CH_2(OH)$), Sodium hydroxide (NaOH), Ammonium hydroxide ($NH_4OH$, 26 % of ammonia) | Co-precipitation | Magnetic $Fe_3O_4$ nanoparticles | Room temp. | 8 nm | Adsorbent, wastewater purification | The adsorption capacity of $Fe_3O_4$ particles increased with decreasing the particle size or increasing the surface area | [184] |
| 20. | Ferric iron (Fe (III), Ferrous iron (II), Sodium borohydrid | Reduction | Zero-valent iron nanoparticles | 23°C | 5 nm - 1000 μm | Separation and transformation of many contaminants | Iron nanoparticles have a core of zero-valent iron and a shell of mainly iron oxides (FeO) | [185] |
| 21. | $FeCl_3 \cdot 6H_2O$ $FeSO_4 \cdot 7H_2O$ NaOH HCl | Co-precipitation | $Fe_3O_4$ magnetic nanoparticles | 80°C | 14 nm cubic | | | [186] |

| Sr. No. | Reagents | Method of preparation | Nature of particles | Temperature | Size and shape | Area of applications | Major outcomes | Ref. |
|---|---|---|---|---|---|---|---|---|
| 22. | Ferric chloride ($FeCl_3 \cdot 6H_2O$) Nickel chloride ($NiCl_2 \cdot 6H_2O$) NaOH Oleic acid | Co-precipitation | Nickel ferrite ($NiFe_2O_4$) nanoparticles | 80°C | 8–28nm inverse spinel structure | Biomedical applications | The super paramagnetic blocking temperature was found to be increasing with increasing particle size that has been attributed to the increased effective anisotropy energy of the nanoparticles | [187] |
| 23. | Ferrous chloride ($FeCl_2 \cdot 4H_2O$) HCl, acetic acid, acetone NaOH | Co-precipitation | $Fe_3O_4$ nanoparticles | 70°C | 10 and 20 nm mono dispersed, spherical-shaped | Hyperthermia therapy | Smaller size, super paramagnetic behaviour at room temperature, higher magnetization value | [188] |
| 24. | $FeCl_2 \cdot 4H_2O$, $FeCl_3 \cdot 6H_2O$, Cyclohexane di-n-propylamine, Ammonia Ethanol | Interfacial co-precipitation | $Fe_3O_4$ nanoparticles | 25°C | 6 – 8 nm spinel | Biomedical | Mechanism of fabricating nanoparticles by interfacial co-precipitation method, and the reaction procedures of the co-precipitation, which should be helpful for the phase control in the preparation of magnetite nanoparticles | [189] |
| 25. | $Fe(NO_3)_3 \cdot 9H_2O$, $M(NO_3)_2$, NaOH Glycine | Hydrothermal | $MFe_2O_4$ (M = Co, Ni, Zn) nanocrystals | 180°C | 30 nm spherical shape | Drug delivery, bio-separation, and magnetic resonance imaging | Prepared magnetic $MFe_2O_4$ nanocrystals are a type characteristic of super paramagnetic materials | [190] |

**Table 5.1.** (Continued)

| Sr. No. | Reagents | Method of preparation | Nature of particles | Temperature | Size and shape | Area of applications | Major outcomes | Ref. |
|---|---|---|---|---|---|---|---|---|
| 26. | Iron bromine (II) and (III) $FeBr_2$, $FeBr_3$, 3-aminopropyl-triethoxysilane (APTES), 3-aminopropyl-ethyl-diethoxysilane (APDES) 3-aminopropyl-diethyl-ethoxysilane (APES) | Sono-precipitation | Magnetite nanoparticles | | 10 nm spherical | Biomedical applications | Colloidal stability is an extremely important parameter for in vivo SPIO biomedical applications and helps to inhibit capillary emboli when carrying biological molecules or drugs | [191] |
| 27. | Iron powders 10 M NaOH | Hydrothermal | Magnetite ($Fe_3O_4$) octahedral particles | 180°C | 500 nm octahedrons | Biomedical applications | The concentration of NaOH and the reaction temperature played a key role in the formation of the magnetite octahedrons | [111] |
| 28. | $Fe(OH)_2$, $Fe(OH)_3$ Ammonium hydroxide Oleic acid (OA) | Co-precipitation | Oleic acid coated $Fe_3O_4$ MNPs | 80°C | < 100 nm 7.8 ± 1.9 nm cubic shape | Biomedical applications | The oleic acid provided the $Fe_3O_4$ particles with better dispersibility | [192] |
| 29. | Iron (III) acetylacetonate (99%), Dibenzylether (99%), Decanoic acid (96%) Hydrazine (anhydrous, 98%), Oleic acid (90%), 1, 2-hexadecanediol (90 %), Oleylamine (70%) | Thermal decomposition | Iron oxide nanoparticles (magnetite/maghemite $Fe_3O_4$ / $\gamma$-$Fe_2O_3$) | 200°C | 4-20 nm stripe-like structures | Biomedical applications | To tune the shape, the size, and the magnetic properties are discussed | [68] |

| Sr. No. | Reagents | Method of preparation | Nature of particles | Temperature | Size and shape | Area of applications | Major outcomes | Ref. |
|---|---|---|---|---|---|---|---|---|
| 30. | Iron chloride ($FeCl_3 \cdot 6H_2O$) Sodium oleate 1-octadecene Ethanol Hexane | Thermal decomposition | Mono disperse nanocrystals (MnO, CoO and Fe) | 320°C | 5 - 22 nm spherical | Biomedical applications | The process allows monodisperse nanocrystals to be obtained on an ultra large scale of 40 g in a single reaction and without a further size-sorting process | [108] |
| 31. | Iron (III) chloride hexahydrate ($FeCl_3 \cdot 6H_2O$) Iron (II) chloride tetrahydrate ($FeCl_2 \cdot 4H_2O$), 2, 2-Diphenyl-1-1-picrylhydrazyl (DPPH) Sodium borohydride Ammonia solution | Ultrasonicated assisted co-precipitation | Silver and iron oxide nanoparticles | Room temp., 60°C | ~20 nm spherical | Biomedical applications | The ultrasound assisted nanoparticles showed higher stability and antibacterial and antioxidant activity compared with the nanoparticles fabricated by magnetic stirring | [193] |
| 32. | Ferric chloride hexahydrate ($FeCl_3 \cdot 6H_2O$), Ferrous chloride tetrahydrate ($FeCl_2 \cdot 4H_2O$), Sodium hydroxide (NaOH), Ammonium hydroxide 25 % ($NH_4OH$) Hydrochloric acid 37% (HCl) | Co-precipitation | Magnetite and maghemite nanoparticles | 95°C | $7.93 \pm 0.08$ nm $9.44 \pm 0.11$ nm $12.77 \pm 0.09$ nm $11.93 \pm 0.13$ nm | Environmental and biomedical applications | The materials obtained from the precipitating agent $NH_4OH$ are more uniform than those obtained with NaOH | [95] |

## Table 5.1. (Continued)

| Sr. No. | Reagents | Method of preparation | Nature of particles | Temperature | Size and shape | Area of applications | Major outcomes | Ref. |
|---|---|---|---|---|---|---|---|---|
| 33. | $Fe^{3+}$, $Fe^{2+}$ $Co^{2+}$ cations in aqueous solution with pH $\approx 7$ NaOH | Co-precipitation | Co-substituted ferrite nanoparticles Co–Fe ferrite nanoparticles | 7°C | 9.5 – 11nm spherical | High-density recording devices, ferrofluids and biomedicine | The Curie temperature $T_C$ and saturation magnetization $M_s$ at nanoscale are lower than those of the bulk and decrease with the increase of cobalt cont. | [73] |
| 34. | Trioctylamine Oleic acid $Fe(CO)_5(CH_3)NO$ | Thermal decomposition | Magnetic iron oxide ($\gamma$-$Fe_2O_3$) nanocrystals | 180°C | 13 nm spherical | Biomedical | The particles stable with respect to aggregation or grain growth and keeps them highly dispersed in a variety of organic medium | [99] |
| 35. | $FeSO_4 \cdot 7H_2O$, $FeCl_3 \cdot 6H_2O$, $NH_4OH$ | Co-precipitation and reverse Co-precipitation | Magnetite nanoparticles | 50°C | 10-15 nm spherical | Drug release, cancer therapy | The involvement of sono-chemical treatment instead of mechanical treatment gave smaller and more homogeneous crystal size | [194] |
| 36. | Ferric chloride hexahydrate ($FeCl_3 \cdot 6H_2O$), Ferrous chloride tetrahydrate ($FeCl_2 \cdot 4H_2O$) Ammonium hydroxide ($NH_4OH$) Polyethylene glycol (PEG-4000) | Chemical co-precipitation | Iron oxide nanoparticles ($Fe_3O_4$) | 80°C | 20-30 nm cubic spinel structure | Biomedical | $Fe_3O_4$ particles prepared are super paramagnetism in nature | [195] |

| Sr. No. | Reagents | Method of preparation | Nature of particles | Temperature | Size and shape | Area of applications | Major outcomes | Ref. |
|---|---|---|---|---|---|---|---|---|
| 37. | Iron (III) chloride hexahydrate ($FeCl_3 \cdot 6H_2O$) Anhydrous sodium sulfate ($Na_2SO_4$) Sodium dihydrogen phosphate dihydrate ($NaH_2PO_4 \cdot 2H_2O$) | Hydrothermal | Iron oxide magnetic short nanotubes | 300°C | 3 to 6 nm, short nanotubes (SNTs) | Ion adsorbents, catalytic fields, biosensors, MRI | The size, morphology, shape, and surface architecture control of the iron oxide SNTs are achieved by simple adjustments of ferric ions concentration without any surfactant addition | [196] |
| 38. | Aqueous $FeCl_2 \cdot 4H_2O$, Aqueous $Na_2SiO_3 \cdot 9H_2O$ | Co-precipitation | Iron oxides and gadolinium nanoparticles | 80°C | 30 nm and 14 nm spherical | Medical application, cancer, HIV, bird flu | Sufficiently high value of saturation magnetization and attraction to magnet ability | [197] |
| 39. | Iron (III) chloride hexahydrate ($FeCl_3 \cdot 6H_2O$) Ammonia hydroxide ($NH_4OH$) Ethanol ($C_2H_6O$) | Co-precipitation Hydrothermal | Pure $\alpha$-$Fe_2O_3$ nanoparticles | 80°C, 160°C | 21 nm and 33 nm rhombohedral (hexagonal) | Electro-magnetic devices, magnetic recording (storage) devices | Pure hematite prepared by hydrothermal method has smallest size, best crystallinity, highest band gap and best value of saturation magnetization compared to the hematite elaborated by the precipitation method | [198] |
| 40. | Ethyl alcohol (EtOH) Iron (III) acetylacetonate $C_{15}H_{12}FeO_6$,Methanol | Sol-gel | Iron oxide $Fe_3O_4$ nanoparticles | NA | 8 nm spherical | Hyperthermia application | Ferromagnetic behaviour at room temperature | [125] |

**Table 5.1. (Continued)**

| Sr. No. | Reagents | Method of preparation | Nature of particles | Temperature | Size and shape | Area of applications | Major outcomes | Ref. |
|---|---|---|---|---|---|---|---|---|
| 41. | Co(acac)$_2$ Fe(acac)$_3$ 1–2 octanediol Oleic acid Oleylaminaoleic acid | Chemical method | CoFe$_2$O$_4$ nanoparticles | 262°C | 2-7 nm Single crystalline | Technological applications | The surface anisotropy is large enough to change the ferrimagnetic order in the particle shell | [199] |
| 42. | Fe(NO$_3$)$_3$·9H$_2$O NaOH, Tergitol NP-9, 1-octanol, cyclohexane, acetone | Emulsion (reverse micelle method) | Iron-oxide nanoparticles | 62°C | < 5 nm and 60 nm amorphous | Biomedical applications | High coercivity iron oxide nanoparticles with high saturation magnetization | [7] |
| 43. | FeCl$_2$·4H$_2$O, FeCl$_3$·6H$_2$O, NaOH, TEOS, EDTA, Glacial acetic acid, Glycerol | Co-precipitation | EDTA-Surface functionalized Si-coated magnetite nanoparticles | 60°C | 12-13 nm spherical | Removal of toxic metals from ground water | Due to superparamagnteic nature the EDTA functionalized Si-coated Fe$_3$O$_4$ nanoparticles loaded with toxic metals can be efficiently separated by applying external magnetic field. Hence it has great potential application in easy and fast separation of toxic metals from waste water | [200] |

The encapsulation of magnetic nanoparticles within a nonpolar phase such as silicon oil or soybean oil may generate oil-in-water emulsions. Sakulchaicharoen et al. [170] prepared FePd particles from $FeCl_2$ using borohydride as the reducing agent, and guar gum as a stabilizer. The guar solution was converted into a gel on addition of borate ions, subsequently the viscosity of guar solution increased from 2.2 to 6.0 cP [150]. This resulted in comparatively good stability to avoid sedimentation of particle within the time range of 48 hours.

# Chapter 6

# Applications of Magnetic Nanoparticles

Major categories of applications of mangeti nanoparticles can be broadly subdevided as in the field of biomedical engineering, and for heat transfer In variety of systems.

## 6.1. Heat Transfer Applications

Even though ambient air is available in abundant, it is not a feasible option for cooling for many technical operations due to it low thermal conductivity. However, water having comparatively high value of specific heat is being used as a coolant for many technical applications. Water as a cooling medium has some limitations, hence in order to improve the heat transfer coefficient, there is a requirement of alternative fluids for cooling applications [201]. On the other hand, solids have higher thermal conductivities than liquids, and offer better thermal properties [202] for heat transfer applications. However, for heat transfer application it is bit difficult use the solid metals directly due to their transportation and handling problems. Use of nanofluids can be the potential candidate for such applications. Nanofluids are the dispersions of nanoparticles in liquid. Particles concentrations have direct effects on the viscosity and thermal conductivity of nanofluids. Nanofluids containing iron oxide nanoparticles are effectively used for such applications. Thermal conductivity is one of the major important properties for the fluid to be used for heat transfer applications.

Even though the liquid metal having a low melting point is the best option as a base fluid for making a superconductive solution, addition of $Fe_3O_4$ magnetic particles significantly enhances its characteristics to be used as a heat transfer fluid. The thermal conductivity of the base fluid can be increased with the addition of more conductive nanoparticles. Stoian and Holotescu [203] experimentally investigated the cooling capabilities of transformer based magnetic nanofluids in the presence of applied magnetic field, and reported significant cooling enhancement due to use of $Fe_3O_4$ magnetic particles. They found that the rate of rising in temperature of the magnetic nanofluid was lower than that of the transformer oil.

## 6.2. Candidate for Next Generation Nuclear Waste Remediation Material

Magnetic nanomaterials exhibit distinct physical properties than that of their bulk counterparts. Magnetic nanoparticles have emerged as a broad class of materials having high surface area and radiation resistance capacity under real time extreme conditions which have high potential towards radioactive metal ions remediation [204]. The radiotoxic metal ions either sit on the cavity of the magnetic nanoparticles or interact with the covalently bonded functional groups which has been found crucial for nuclear waste remediation especially for sequestration of radiotoxic materials including actinides and lanthanides [204].

## 6.3. Magnetic Nanoparticles for Wastewater Treatment

Magnetic iron oxide nanoparticles have been reported [205] for their efficacy for removal of phosphate in sewage wastewater. Water-in-oil microemulsion was used to prepare magnetic iron oxide nanoparticles having size range 7 to 10 nm. Lakshamanan et al. [205] achieved more than 95% of phosphate removal in 5 min, and around 100% of phosphate removal in 20 min from sewage wastewater.

## 6.4. Cooling in Nuclear Power Plants

Following are the major cooling requirements in nuclear power plants:

- Pressurized water reactors
- Pressurized water reactors and boiling water reactors
- Maintaining molten core inside the vessel after catastrophic accidents in high-power density light water reactors

Nanofluids are used as the primary reactor coolant in the pressurized water reactors This helps improving power uprates leading to significantly enhance their economic performance. Emergency core cooling systems are very crucial for pressurized water reactors and boiling water reactors and boiling water reactors.

## 6.5. Applications in the Space and Defence Sectors

Efficient cooling systems are in the great demand for applications in the space, aviation and defence sectors due to constraints of small volume and light weight. Cooling systems offering high heat transfer coefficient with compact design are of much importance for such applications. Many military equipments need high-heat flux cooling system requirements which can ensure their reliable functioning. Military vehicles, submarines, high-power laser diodes, etc. may benefit from the cooling provided by nanofluids with high critical heat fluxes. Since power density is extremely high in the space and defence industries; smaller and lighter components designs are required, and nanofluids find extensive use in such cooling applications.

## 6.6. Other Thermal Applications

Ming et al. [206] experimentally studied the effects of magnetic nanofluids as the working fluid on the performance of a flat plate heat pipe. The external magnetic field was found to be capable for enhancing the turbulence and recirculation of the magnetic fluid, leading to increased heat transfer coefficient. More uniform heat flux distribution on the condensation surface can be ensured for the flat plate heat pipe with the circulation of magnetic fluid than that achieved by water as a heat transfer media [207].

Chiang et al. [208] demonstrated the closed loop heat pipe using magnetic nanofluids which improved the thermal properties. The results showed that optimal thermal conductivities achieved with the applied field of 200 Oe. Their results confirmed that magnetically enhanced heat transfer can be easily applied for heat removal in the device or the energy systems.

Taslimifar et al. [209] experimentally evaluated the thermal performance of open loop pulsating heat pipes using magnetic nanoparticles based ferrofluid. It was found that the thermal performance of heat pipe with ferrofluid improved under the influence of the magnetic field. The magnetic particle based thermal conversion is advantageous for open loop pulsating heat pipes.

Fumoto et al. [210] studied heat transport in miniature devices using temperature sensitive magnetic nanofluids. The effect of parameters like a magnetic force, the position of magnet and the fluid temperature were experimentally evaluated. These have predominant effect on the dynamics of the magnetic nanofluids. It was found that the magnetic nanofluids can be controlled by changing the magnetic force and position of the magnet.

Philip et al. [211] designed a multipurpose device in which magnetic nanofluids were used. They evaluated the ratio of thermal conductivity to viscosity and tested for the heat dissipater and damper. This was found useful for both heat dissipater and damper. Implementation of such design concepts for certain microfluidic systems and microelectromechanical systems can really results in more efficient performance of the devices.

## 6.7. Cooling of Electronics Microchips

Due to the innovations in automations many devices nowadays are running with the precise control by electronics chips or microchips. The cooling of these chips or electronic circuitries is very crucial for effective operation of the electronics systems. Rapid dissipation of the heat from such electronics devices are of prime importance for their appropriate functioning and to avoid possible damage to the devices. Magnetic nanofluids can be effectively used for cooling of microprocessor due to their high thermal conductivity.

## 6.8. Microfluidics Applications

Separation of microparticles or nanoparticles is very difficult due to the confined region of access, tiny quantities of the feed and quantity or the flow rate of throughput. Use of the external magnetic field in addition to the dispersion of nanofluid in to the feed material may cause certain degree of separation based on the geometry of the microchannels and orientation of the magnetic field [212, 213].

Accordingly, the concentration of the streams flowing in microchannels can be controlled by selective diverting of the particles treated with magnetical nanofluids. Appropriate orientation of the magnetic field, and the intensity of magnetic field can govern the bifurcation of the flow path of the particles treated with magnetic nanofluids [214].

The scrupulous handling of a small amount of liquid is necessary for fluidic digital display devices, optical devices and microelectromechanical systems using lab-on-a-chip analysis systems [202]. Electrowetting and surface characteristics of various surfaces can be characterized by the contact angle measurements. Variation of contact angle through applied voltage can be precisely controlled with the small volume of the liquid sample. Electrowetting with a dielectric is the most versatile method incorporating small amount of liquid sample. Nanofluids are very effective for surface treatment of different substrates.

## 6.9. Mechanical Applications

Tribological applications often requires the surface smoothening in order to avoid mechanical wear or excessive friction. Magnetic nanofluids can be effectively used for lubricating the surfaces and guiding the nanoparticles in certain direction depending on the friction experienced by the surfaces. Nanofluids have exceptional lubricating characteristic because nanoparticles form protective films with low hardness and elastic modulus, and prevent the substrates from excessive friction or surface damage.

Magnetic liquid rotary seals function uniquely and used for a very wide range of applications with minimal maintenance and exceptionally minimal leakages. These can be facilitated by utilising the magnetic characteristics of the liquid-bound magnetic nanoparticles for specific applications.

## 6.10. Magnetic Sealing

Ferromagnetic nanoparticles suspended in base fluids are the special class of suspensions used for magnetic sealing. They ate stable colloidal suspension of small magnetic nanoparticles like magnetite. Large variety of industrial rotating equipments operated at high rotating speed require effective mechanical sealing in order to avoid the leakage of harmful fluids. Magnetic sealing provides allow-cost solution for environmental-friendly hazardous-gas sealing with low frictional losses with long life and high-reliability [215]. A ring-shaped magnet is used as a part of a magnetic system in which an intense magnetic field is applied in the gaps between the magnetically permeable shaft and the surface of an opposing pole block. Magnetic nanofluid is injected into the gaps to create discrete liquid rings in order to support a pressure difference. This ensures zero leakage around he moving part. The sealing system is designed so as to not coming in contact close contact with the moving parts or in other words, it does not touch any moving parts during the operation. Therefore, sealing fluids with magnetic liquids can be easily used in such application areas. Kim et al. [216] showed the efficacy of magnetic particles dispersed in nanofluids for the sealing in the rotary vacuum pump.

## 6.11. Biomedical Applications

Magnetic nanoparticles and nanofluids are widely used for variety of applications in the area of biomedical usage, such as magnetic drug delivery, hyperthermia therapy, and magnetic separation of cells.

Magnetic iron oxide nanoparticles have been reported for their applications in disease detection as well as therapy [217]. Magnetic iron oxide nanoparticles are capable to generate imaging contrast. One of the major reasons for the versatile use of magnetic iron oxide nanoparticles is that they provide mechanical and thermal energy in vivo in response to an external magnetic field, which make them useful for clinical interventions. Magnetite and maghemite nanocrystals support only one magnetic domain at the length scale less than the magnetic domain wall width of approximately 80-90 nm. The thermal fluctuation becomes large enough to overcome the energy barrier due to magnetic anisotropy at the nanocrystal size less than 20 nm which is the superparamagnetic limit (Krishnan et al., 2006). Saturation magnetization is reached by the nanocrystals when exposed to the modest magnetic field which makes them to known as superparamagnetic iron oxide nanoparticles. Due to their innocuous toxicity profile and biocompatibility, superparamagnetic iron oxide nanoparticles have been successfully used as the contrast agent for magnetic resonance imaging. Reactive surface of the magnetic nanoparticles can be advantageously used for biocompatible coatings for certain applications [218].

The intrinsic properties of magnetic nanoparticles provide a non-invasive means for controlling the fate for many applications of biomedically importance [218], specifically

- Magnetic resonance imaging
- Magnetic drug delivery

- Magnetic gene delivery
- Magnetic fluid hyperthermia
- Magnetic separation
- Biosensors

## 6.12. Magnetic Drug Delivery

The important role of the magnetic nanofluids in the biomedical field is to reduce the side effects of radiation during cancer therapy or other diseases. Iron-based magnetic nanoparticles can be useful in delivering drugs or radiation to the targeted cells without rapturing the healthy cells [219].

Nakano et al. [220] proposed the drug delivery system using magnetic nanofluids. This drug delivery system incorporates a ferrofluid cluster consist of magnetic nanoparticles. Such a ferrofluid subjected to the drug is injected to the cancer tumour where it is kept for an hour through a well-focused magnetic field. The volume of the drug required for this targeted delivery is much less than that of otherwise dispersed in the whole body. As the magnetic field is called off the drug will be dispersed in the whole body but as the total amount is very less there will be no side effects as reported in the literature [221].

Magnetic nanofluids also improve cooling around the surgical area, hence it ensures the appropriate healing and prevent the danger of organ damage.

## 6.13. Hyperthermia Therapy

Hyperthermia is mostly characterised by the increase of abnormal body temperature artificially made by external medical devices. Thus, the symptoms are completely different from fever and heatstroke, which need to be controlled by the set point of the body temperature. Dispersed magnetic nanoparticles in nanofluids used as a mediator to accomplish the cancer treatment with higher efficiency. In this method, the mediated particles are injected within the affected area in the form of colloidal suspension under the influence of magnetic force. It offers the elimination of cancer cells through the formatted heat of suspended particles by magnetic force without disturbing the nearby unaffected cells. The speciality of magnetic nanofluids is that it can enhance the local temperature of the affected part with the inclusion of magnetic nanoparticles to cancer cells. The increased temperature of magnetic nanoparticles promotes hyperthermia therapy [222].

# Chapter 7

# Concluding Remarks and Perspectives

Remarkable work has been done in the area of nanosynthesis during the last couple of decades. Iron oxide nanoparticles have been successfully implemented in vivo/in vitro biomedical field. Different types of monodisperse nanomaterials with controlled size and shape have been synthesized using several chemical methods. However, synthesis of iron oxide nanoparticles still requires more development in synthesis methods especially leading to process intensification for the magnetic nanoparticle synthesis. It is necessary to understand the chemistry of coated surface-engineered pathways, reaction mechanism, stability, and the synthesis of particles of uniform size and shape. Surface modification of particles depends on the type of surface treatment, methodology of coating, and types of coating materials such as polymer coating, small molecular coating, silica coating, metal coating and liposome coating [171]. However, the coatings of functional materials have different film thickness due to their coupling effect can satisfactorily enlarge the hydrodynamic size. Consequently, enlarged hydrodynamic size alter the surface properties of iron oxide nanoparticles. Thus, it is necessary to measure the hydrodynamic size and zeta potential [15]. The shape of particles is required to measure because of several investigations have shown its effects biomedical research [146].

Moreover, a big challenge is to overcome large tendency to aggregate and week magnetization of magnetic nanoparticles. Generally, the choice of synthesis method is not a big factor for production of particles at lab scale. For large-scale production or industrial point of view, precipitation method is preferred over other synthesis methods because precipitation process is comparatively easily practically scalable. Therefore, it is necessary to explore the synthesis method for further enhancement of the physical properties of magnetic nanoparticles.

Synthesis of iron oxide nanoparticles require addition of expensive precursors or toxic chemicals, and most of reactions carried out in an organic phase at high temperature and require large dilution. In most of the cases, synthesised nanoparticles are dispersible in organic solvents, instead of aqueous phase. Therefore, there is a requirement of more convenient synthesis method to produce water soluble metal oxide or particularly metallic nanoparticles having controlled size and shape [104].

The saturation magnetization of uncoated and modified iron oxide nanoparticles is the parameter that describes the magnetic behaviour of magnetic nanoparticles. The saturation magnetization decreases if magnetic nanoparticles are solidity stabilized, and it increases with agglomeration of nanoparticles.

The surface functionalization and modification is found to improve functionality of magnetic nanoparticles. For quality research, selection of better and faster methods which improve properties of magnetic nanoparticles should be considered for advancement in this active field [172].

Process intensification, conceptual design and scale-up will provide a unique platform for advancement in the design, synthesis, surface engineering, and biocompatibility of nanomaterials in near future. Further research and innovative approach to enhance the efficacy of magnetic iron oxide nanoparticles for industrial applications including biomedical usage will broaden the scope of applications in day-to-day life as well as commercial demands.

# References

[1] Lee, H., E. Lee, D. K. Kim, N. K. Jang, Y. Y. Jeong, S. Jon, Antibiofouling Polymer-Coated Superparamagnetic Iron Oxide Nanoparticles as Potential Magnetic Resonance Contrast Agents for *in Vivo* Cancer Imaging, *J Am Chem Soc.* 128 (2006) 7383–7389. https://doi.org/10.1021/ja061529k.

[2] Jun, Y., J. Seo, J. Cheon, Nanoscaling Laws of Magnetic Nanoparticles and Their Applicabilities in Biomedical Sciences, *Acc Chem Res.* 41 (2008) 179–189. https://doi.org/10.1021/ar700121f.

[3] Wu, W., C. Z. Jiang, V. A. L. Roy, Designed synthesis and surface engineering strategies of magnetic iron oxide nanoparticles for biomedical applications, *Nanoscale.* 8 (2016) 19421–19474. https://doi.org/10.1039/C6NR07542H.

[4] Cornell, R. M., U. Schwertmann, *The iron oxides: structure, properties, reactions, occurrences, and uses*, 2nd, compl ed., Weinheim : Wiley-VCH, 2003.

[5] Tuček, J., R. Zbořil, A. Namai, S. Ohkoshi, ε-$Fe_2O_3$: An Advanced Nanomaterial Exhibiting Giant Coercive Field, Millimeter-Wave Ferromagnetic Resonance, and Magnetoelectric Coupling, *Chemistry of Materials.* 22 (2010) 6483–6505. https://doi.org/10.1021/cm101967h.

[6] Cui, H., Y. Liu, W. Ren, Structure switch between α-$Fe_2O_3$, γ-$Fe_2O_3$ and $Fe_3O_4$ during the large scale and low temperature sol–gel synthesis of nearly monodispersed iron oxide nanoparticles, *Advanced Powder Technology.* 24 (2013) 93–97. https://doi.org/10.1016/j.apt.2012.03.001.

[7] Dar, M. I., S. A. Shivashankar, Single crystalline magnetite, maghemite, and hematite nanoparticles with rich coercivity, *RSC Adv.* 4 (2014) 4105–4113.

[8] Sakurai, S., A. Namai, K. Hashimoto, S. Ohkoshi, First Observation of Phase Transformation of All Four $Fe_2O_3$ Phases (γ → ε → β → α-Phase), *J Am Chem Soc.* 131 (2009) 18299–18303. https://doi.org/10.1021/ja9046069.

[9] Klahr, B. M., A. B. F. Martinson, T.W. Hamann, Photoelectrochemical Investigation of Ultrathin Film Iron Oxide Solar Cells Prepared by Atomic Layer Deposition, *Langmuir.* 27 (2011) 461–468. https://doi.org/10.1021/la103541n.

[10] Saremi-Yarahmadi, S., K. G. U. Wijayantha, A.A. Tahir, B. Vaidhyanathan, Nanostructured α-$Fe_2O_3$ Electrodes for Solar Driven Water Splitting: Effect of Doping Agents on Preparation and Performance, *The Journal of Physical Chemistry C.* 113 (2009) 4768–4778. https://doi.org/10.1021/jp808453z.

[11] Wu, C., P. Yin, X. Zhu, C. OuYang, Y. Xie, Synthesis of Hematite (α-$Fe_2O_3$) Nanorods: Diameter-Size and Shape Effects on Their Applications in Magnetism,

Lithium Ion Battery, and Gas Sensors, *J Phys Chem B*. 110 (2006) 17806–17812. https://doi.org/10.1021/jp0633906.

[12] Zeng, S., K. Tang, T. Li, Z. Liang, D. Wang, Y. Wang, W. Zhou, Hematite Hollow Spindles and Microspheres: Selective Synthesis, Growth Mechanisms, and Application in Lithium Ion Battery and Water Treatment, *The Journal of Physical Chemistry* C. 111 (2007) 10217–10225. https://doi.org/10.1021/jp0719661.

[13] Zhang, G., Y. Gao, Y. Zhang, Y. Guo, $Fe_2O_3$-Pillared Rectorite as an Efficient and Stable Fenton-Like Heterogeneous Catalyst for Photodegradation of Organic Contaminants, *Environ Sci Technol*. 44 (2010) 6384–6389. https://doi.org/10.1021/es1011093.

[14] Cheng, C. J., C. C. Lin, R. K. Chiang, C. R. Lin, I. S. Lyubutin, E. A. Alkaev, H. Y. Lai, Synthesis of monodisperse magnetic iron oxide nanoparticles from submicrometer hematite powders, *Cryst Growth Des*. 8 (2008) 877–883. https://doi.org/10.1021/cg0706013.

[15] Wu, W., Z. Wu, T. Yu, C. Jiang, W.-S. Kim, Recent progress on magnetic iron oxide nanoparticles: synthesis, surface functional strategies and biomedical applications, *Sci Technol Adv Mater*. 16 (2015) 23501. https://doi.org/10.1088/1468-6996/16/2/023501.

[16] Zboril, R., M. Mashlan, D. Petridis, Iron(III) Oxides from Thermal ProcessesSynthesis, Structural and Magnetic Properties, Mössbauer Spectroscopy Characterization, and Applications, *Chemistry of Materials*. 14 (2002) 969–982. https://doi.org/10.1021/cm0111074.

[17] Morin, F. J. Magnetic Susceptibility of Magnetic Susceptibility of αFe2O3 and α-$Fe_2O_3$ with Added Titanium with Added Titanium, *Physical Review*. 78 (1950) 819–820. https://doi.org/10.1103/PhysRev.78.819.2.

[18] Artman, J. O., J. C. Murphy, S. Foner, Magnetic Anisotropy in Antiferromagnetic Corundum-Type Sesquioxides, *Physical Review*. 138 (1965) A912–A917. https://doi.org/10.1103/PhysRev.138.A912.

[19] Moriya, T. Anisotropic Superexchange Interaction and Weak Ferromagnetism, *Physical Review*. 120 (1960) 91–98. https://doi.org/10.1103/PhysRev.120.91.

[20] Dzyaloshinsky, I., A thermodynamic theory of "weak" ferromagnetism of antiferromagnetics, *Journal of Physics and Chemistry of Solids*. 4 (1958) 241–255. https://doi.org/https://doi.org/10.1016/0022-3697(58)90076-3.

[21] Hill, R. J., J. R. Craig, G. V Gibbs, Systematics of the spinel structure type, *Phys Chem Miner*. 4 (1979) 317–339. https://doi.org/10.1007/BF00307535.

[22] Gossuin, Y., P. Gillis, A. Hocq, Q. L. Vuong, A. Roch, Magnetic resonance relaxation properties of superparamagnetic particles, *Wiley Interdiscip Rev Nanomed Nanobiotechnol*. 1 (2009) 299–310. https://doi.org/10.1002/wnan.36.

[23] Shafi, K. V. P. M., A. Ulman, X. Yan, N. L. Yang, C. Estournès, H. White, M. Rafailovich, Sonochemical Synthesis of Functionalized Amorphous Iron Oxide Nanoparticles, *Langmuir*. 17 (2001) 5093–5097. https://doi.org/10.1021/la010421+.

[24] Paul, K. Magnetic and transport properties of monocrystalline $Fe_3O_4$, *Open Physics*. 3 (n.d.) 115–126. https://doi.org/https://doi.org/10.2478/BF02476510.

[25] Munir, A. *Magnetic Nanoparticle Enhanced Actuation Strategy for mixing, separation, and detection of biomolecules in a Microfluidic Lab-on-a-Chip System.* : Worcester Polytechnic Institute, (2012).

[26] Dar, M. I., S. A. Shivashankar, Single crystalline magnetite, maghemite, and hematite nanoparticles with rich coercivity, *RSC Adv.* 4 (2014) 4105–4113. https://doi.org/10.1039/C3RA45457F.

[27] Dronskowski, R. The little maghemite story: A classic functional material, *Adv Funct Mater.* 11 (2001) 27–29. https://doi.org/10.1002/1616-3028(200102)11:1%3C27::AID-ADFM27%3E3.0.CO;2-X.

[28] Lévy, M., C. Wilhelm, J. M. Siaugue, O. Horner, J. C. Bacri, F. Gazeau, Magnetically induced hyperthermia: size-dependent heating power of γ-Fe$_2$O$_3$ nanoparticles, *Journal of Physics: Condensed Matter.* 20 (2008) 204133. https://doi.org/10.1088/0953-8984/20/20/204133.

[29] Wang, W. W., Y. J. Zhu, M. L. Ruan, Microwave-assisted synthesis and magnetic property of magnetite and hematite nanoparticles, *Journal of Nanoparticle Research.* 9 (2007) 419–426. https://doi.org/10.1007/s11051-005-9051-8.

[30] Chin, S. F., M. Makha, C. L. Raston, M. Saunders, Magnetite ferrofluids stabilized by sulfonato-calixarenes, *Chemical Communications.* (2007) 1948–1950. https://doi.org/10.1039/B618596G.

[31] Wan, J., W. Cai, X. Meng, E. Liu, Monodisperse water-soluble magnetite nanoparticles prepared by polyol process for high-performance magnetic resonance imaging, *Chemical Communications.* (2007) 5004–5006. https://doi.org/10.1039/B712795B.

[32] Zayat, M., F. del Monte, M. P. Morales, G. Rosa, H. Guerrero, C. J. Serna, D. Levy, Highly Transparent γ-Fe2O3/Vycor-Glass Magnetic Nanocomposites Exhibiting Faraday Rotation, *Advanced Materials.* 15 (2003) 1809–1812. https://doi.org/10.1002/adma.200305436.

[33] Pinho, S. L. C., G. A. Pereira, P. Voisin, J. Kassem, V. Bouchaud, L. Etienne, J. A. Peters, L. Carlos, S. Mornet, C. F. G. C. Geraldes, J. Rocha, M.-H. Delville, Fine Tuning of the Relaxometry of γ-Fe$_2$O$_3$@SiO$_2$ Nanoparticles by Tweaking the Silica Coating Thickness, *ACS Nano.* 4 (2010) 5339–5349. https://doi.org/10.1021/nn101129r.

[34] Wang, B., J. Hai, Q. Wang, T. Li, Z. Yang, Coupling of Luminescent Terbium Complexes to Fe$_3$O$_4$ Nanoparticles for Imaging Applications, *Angewandte Chemie International Edition.* 50 (2011) 3063–3066. https://doi.org/10.1002/anie.201006195.

[35] Petrova, O., E. Gudilin, A. Chekanova, A. Knot'ko, G. Murav'eva, Y. Maksimov, V. Imshennik, I. Suzdalev, Y. Tretyakov, Microemulsion synthesis of mesoporous γ-Fe$_2$O$_3$ nanoparticles, *Doklady Chemistry* - DOKL CHEM. 410 (2006) 174–177. https://doi.org/10.1134/S001250080610003X.

[36] Jing, Z., Y. Wang, S. Wu, Preparation and gas sensing properties of pure and doped γ-Fe$_2$O$_3$ by an anhydrous solvent method, *Sens Actuators B Chem.* 113 (2006) 177–181. https://doi.org/https://doi.org/10.1016/j.snb.2005.02.045.

[37] Vidal-Vidal, J., J. Rivas, M. A. López-Quintela, Synthesis of monodisperse maghemite nanoparticles by the microemulsion method, *Colloids Surf A*

## References

*Physicochem Eng Asp.* 288 (2006) 44–51. https://doi.org/10.1016/j.colsurfa.2006.04.027.

[38] Du, G. H., Z. L. Liu, X. Xia, Q. Chu, S. M. Zhang, Characterization and application of Fe3O4/SiO2 nanocomposites, *J Solgel Sci Technol.* 39 (2006) 285–291. https://doi.org/10.1007/s10971-006-7780-5.

[39] Hyeon, T., S. S. Lee, J. Park, Y. Chung, H. Bin Na, Synthesis of Highly Crystalline and Monodisperse Maghemite Nanocrystallites without a Size-Selection Process, *J Am Chem Soc.* 123 (2001) 12798–12801. https://doi.org/10.1021/ja016812s.

[40] Goya, G. F., M. Veith, R. Rapalavicuite, H. Shen, S. Mathur, Thermal hysteresis of spin reorientation at Morin transition in alkoxide derived hematite nanoparticles, *Applied Physics A.* 80 (2005) 1523–1526. https://doi.org/10.1007/s00339-003-2381-4.

[41] Janardhanan, S. K., I. Ramasamy, B. U. Nair, Synthesis of iron oxide nanoparticles using chitosan and starch templates, *Transition Metal Chemistry.* 33 (2008) 127–131. https://doi.org/10.1007/s11243-007-9033-z.

[42] He, Y. T., J. Wan, T. Tokunaga, Kinetic stability of hematite nanoparticles: the effect of particle sizes, *Journal of Nanoparticle Research.* 10 (2008) 321–332. https://doi.org/10.1007/s11051-007-9255-1.

[43] Jones, F., M. I. Ogden, A. Oliveira, G. M. Parkinson, W. R. Richmond, The effect of phosphonate-based growth modifiers on the morphology of hematite nanoparticles formed via acid hydrolysis of ferric chloride solutions, *Cryst Eng Comm.* 5 (2003) 159–163. https://doi.org/10.1039/B302911E.

[44] Herrera, A. P., M. Rodríguez, M. Torres-Lugo, C. Rinaldi, Multifunctional magnetite nanoparticles coated with fluorescent thermo-responsive polymeric shells, *J Mater Chem.* 18 (2008) 855–858. https://doi.org/10.1039/B718210D.

[45] Bilecka, I., I. Djerdj, M. Niederberger, One-minute synthesis of crystalline binary and ternary metal oxide nanoparticles, *Chemical Communications.* (2008) 886–888. https://doi.org/10.1039/B717334B.

[46] Wu, W., Q. He, C. Jiang, Magnetic iron oxide nanoparticles: synthesis and surface functionalization strategies, *Nanoscale Res Lett.* 3 (2008) 397. https://doi.org/10.1007/s11671-008-9174-9.

[47] Griffiths, D. *Introduction to elementary particles*, John Wiley & Sons, 2008.

[48] Nguyen, N. T. Micro-magnetofluidics: interactions between magnetism and fluid flow on the microscale, *Microfluid Nanofluidics.* 12 (2012) 1–16. https://doi.org/10.1007/s10404-011-0903-5.

[49] Pamme, N. Magnetism and microfluidics, *Lab Chip.* 6 (2006) 24–38. https://doi.org/10.1039/B513005K.

[50] Singamaneni, S., V. N. Bliznyuk, C. Binek, E. Y. Tsymbal, Magnetic nanoparticles: recent advances in synthesis, self-assembly and applications, *J Mater Chem.* 21 (2011) 16819–16845. https://doi.org/10.1039/C1JM11845E.

[51] Frenkel, J., J. Doefman, Spontaneous and Induced Magnetisation in Ferromagnetic Bodies, *Nature.* 126 (1930) 274–275. https://doi.org/10.1038/126274a0.

[52] Rikken, R. S. M., R. J. M. Nolte, J. C. Maan, J. C. M. van Hest, D. A. Wilson, P. C. M. Christianen, Manipulation of micro-and nanostructure motion with magnetic fields, *Soft Matter.* 10 (2014) 1295–1308. https://doi.org/10.1039/C3SM52294F.

[53] Néel, L. Antiferromagnetism and Ferrimagnetism, *Proceedings of the Physical Society. Section A.* 65 (1952) 869–885. https://doi.org/10.1088/0370-1298/65/11/301.

[54] Leslie-Pelecky, D. L., R. D. Rieke, Magnetic Properties of Nanostructured Materials, *Chemistry of Materials.* 8 (1996) 1770–1783. https://doi.org/10.1021/cm960077f.

[55] Diamagnetic and Paramagnetic Susceptibilities, *The Physical Principles of Magnetism.* (2001) 31–77. https://doi.org/doi:10.1002/9780470546581.ch2.

[56] Akbarzadeh, A., M. Samiei, S. Davaran, Magnetic nanoparticles: preparation, physical properties, and applications in biomedicine, *Nanoscale Res Lett*, 7 (2012) 144. https://doi.org/10.1186/1556-276X-7-144.

[57] Bennett, L. H., C. H. Page, L. J. Swartzendruber, *Comments on Units in Magnetism*, *J Res Natl Bur Stand.* 83 (1977) 9-12. https://doi.org/ 10.6028/jres.083.002.

[58] Elliott, R. J., A. F. Gibson, E. v. Mielczarek, *An Introduction to Solid State Physics and its Applications*, *Phys Today.* 28 (1975) 58–58. https://doi.org/10.1063/1.3068971.

[59] LaConte, L., N. Nitin, G. Bao, Magnetic nanoparticle probes, *Materials Today.* 8 (2005) 32–38. https://doi.org/https://doi.org/10.1016/S1369-7021(05)00893-X.

[60] Hyeon, T. Chemical synthesis of magnetic nanoparticles, *Chemical Communications.* (2003) 927–934. https://doi.org/10.1039/B207789B.

[61] Park, J., E. Lee, N. M. Hwang, M. Kang, S. C. Kim, Y. Hwang, J. G. Park, H. J. Noh, J. Y. Kim, J. H. Park, T. Hyeon, One-Nanometer-Scale Size-Controlled Synthesis of Monodisperse Magnetic Iron Oxide Nanoparticles, *Angewandte Chemie International Edition.* 44 (2005) 2872–2877. https://doi.org/10.1002/anie.200461665.

[62] Sun, S., H. Zeng, Size-Controlled Synthesis of Magnetite Nanoparticles, *J Am Chem Soc.* 124 (2002) 8204–8205. https://doi.org/10.1021/ja026501x.

[63] Sun, S., H. Zeng, D. B. Robinson, S. Raoux, P. M. Rice, S. X. Wang, G. Li, Monodisperse $MFe_2O_4$ (M = Fe, Co, Mn) Nanoparticles, *J Am Chem Soc.* 126 (2004) 273–279. https://doi.org/10.1021/ja0380852.

[64] Yu, W. W., J. C. Falkner, C. T. Yavuz, V. L. Colvin, Synthesis of monodisperse iron oxide nanocrystals by thermal decomposition of iron carboxylate salts, *Chemical Communications.* (2004) 2306–2307. https://doi.org/10.1039/B409601K.

[65] Ayyub, P., M. Multani, M. Barma, V. R. Palkar, R. Vijayaraghavan, Size-induced structural phase transitions and hyperfine properties of microcrystalline $Fe_2O_3$, *Journal of Physics C: Solid State Physics.* 21 (1988) 2229–2245. https://doi.org/10.1088/0022-3719/21/11/014.

[66] Jolivet, J. P., E. Tronc, C. Chanéac, Iron oxides: From molecular clusters to solid. A nice example of chemical versatility, *Comptes Rendus Geoscience.* 338 (2006) 488–497. https://doi.org/https://doi.org/10.1016/j.crte.2006.04.014.

[67] Sakurai, S., K. Tomita, K. Hashimoto, H. Yashiro, S. Ohkoshi, Preparation of the Nanowire Form of $\varepsilon$-$Fe_2O_3$ Single Crystal and a Study of the Formation Process, *The Journal of Physical Chemistry C.* 112 (2008) 20212–20216. https://doi.org/10.1021/jp806336f.

# References

[68] Guardia, P., A. Labarta, X. Batlle, Tuning the Size, the Shape, and the Magnetic Properties of Iron Oxide Nanoparticles, *The Journal of Physical Chemistry C.* 115 (2011) 390–396. https://doi.org/10.1021/jp1084982.

[69] Chavali, M. S., M. P. Nikolova, Metal oxide nanoparticles and their applications in nanotechnology, *SN Appl Sci.* 1 (2019). https://doi.org/10.1007/s42452-019-0592-3.

[70] Huynh, K. H., X. H. Pham, J. Kim, S. H. Lee, H. Chang, W. Y. Rho, B. H. Jun, Synthesis, properties, and biological applications of metallic alloy nanoparticles, *Int J Mol Sci.* 21 (2020) 1–29. https://doi.org/10.3390/ijms21145174.

[71] Niederberger, M. Nonaqueous sol–gel routes to metal oxide nanoparticles, *Acc Chem Res.* 40 (2007) 793–800. https://doi.org/10.1021/ar600035e.

[72] Ali, R., A. Mahmood, M. A. Khan, A. H. Chughtai, M. Shahid, I. Shakir, M. F. Warsi, Impacts of Ni–Co substitution on the structural, magnetic and dielectric properties of magnesium nano-ferrites fabricated by micro-emulsion method, *J Alloys Compd.* 584 (2014) 363–368. https://doi.org/10.1016/j.jallcom.2013.08.114.

[73] Lan, N. T., N. P. Duong, T. D. Hien, Influences of cobalt substitution and size effects on magnetic properties of coprecipitated Co–Fe ferrite nanoparticles, *J Alloys Compd.* 509 (2011) 5919–5925. https://doi.org/10.1016/j.jallcom.2011.03.050.

[74] Cabanas, A., M. Poliakoff, The continuous hydrothermal synthesis of nanoparticulate ferrites in near critical and supercritical water, *J Mater Chem.* 11 (2001) 1408–1416. https://doi.org/10.1039/B009428P.

[75] Chomoucka, J., J. Drbohlavova, D. Huska, V. Adam, R. Kizek, J. Hubalek, Magnetic nanoparticles and targeted drug delivering, *Pharmacol Res.* 62 (2010) 144–149. https://doi.org/10.1016/j.phrs.2010.01.014.

[76] Tailhades, P., L. Bouet, L. Presmanes, A. Rousset, Thin films and fine powders of ferrites: Materials for magneto-optical recording media, *Le Journal de Physique IV.* 7 (1997) C1-691. https://doi.org/10.1051/jp4:19971283.

[77] Edmundson, M. C., M. Capeness, L. Horsfall, Exploring the potential of metallic nanoparticles within synthetic biology, *N Biotechnol.* 31 (2014) 572–578. https://doi.org/10.1016/j.nbt.2014.03.004.

[78] Shah, M., D. Fawcett, S. Sharma, S. Tripathy, G. Poinern, Green synthesis of metallic nanoparticles via biological entities, *Materials.* 8 (2015) 7278–7308. https://doi.org/10.3390/ma8115377.

[79] Pantidos, N., L. E. Horsfall, Biological synthesis of metallic nanoparticles by bacteria, fungi and plants, *J Nanomed Nanotechnol.* 5 (2014) 1. https://doi.org/10.4172/2157-7439.1000233.

[80] Lin, X. M., C. M. Sorensen, K. J. Klabunde, G. C. Hajipanayis, Control of cobalt nanoparticle size by the germ-growth method in inverse micelle system: size-dependent magnetic properties, *J Mater Res.* 14 (1999) 1542–1547. https://doi.org/10.1557/JMR.1999.0207.

[81] Petit, C., A. Taleb, M. P. Pileni, Cobalt nanosized particles organized in a 2D superlattice: synthesis, characterization, and magnetic properties, *J Phys Chem B.* 103 (1999) 1805–1810. https://doi.org/10.1021/jp982755m.

## References

[82] McNamara, K., S. A. M. Tofail, Nanosystems: the use of nanoalloys, metallic, bimetallic, and magnetic nanoparticles in biomedical applications, *Physical Chemistry Chemical Physics.* 17 (2015) 27981–27995. https://doi.org/10.1039/C5CP00831J.

[83] Lu, L. Y., L. N. Yu, X. G. Xu, Y. Jiang, Monodisperse magnetic metallic nanoparticles: synthesis, performance enhancement, and advanced applications, *Rare Metals.* 32 (2013) 323–331. https://doi.org/10.1007/s12598-013-0117-y.

[84] Hergt, R., S. Dutz, R. Müller, M. Zeisberger, Magnetic particle hyperthermia: nanoparticle magnetism and materials development for cancer therapy, *Journal of Physics: Condensed Matter.* 18 (2006) S2919. https://doi.10.1088/0953-8984/18/38/S26.

[85] Wu, A., X. Yang, H. Yang, Magnetic properties of carbon-encapsulated Fe–Co alloy nanoparticles, *Dalton Transactions.* 42 (2013) 4978–4984. https://doi.org/10.1039/C3DT32639J.

[86] Carta, D., G. Mountjoy, M. Gass, G. Navarra, M. F. Casula, A. Corrias, Structural characterization study of FeCo alloy nanoparticles in a highly porous aerogel silica matrix, *J Chem Phys.* 127 (2007) 204705. https://doi.org/10.1063/1.2799995.

[87] Nguyen, Q., C. N. Chinnasamy, S. D. Yoon, S. Sivasubramanian, T. Sakai, A. Baraskar, S. Mukerjee, C. Vittoria, V. G. Harris, Functionalization of FeCo alloy nanoparticles with highly dielectric amorphous oxide coatings, *J Appl Phys.* 103 (2008) 07D532. https://doi.org/10.1063/1.2839593.

[88] Wu, H., C. Qian, Y. Cao, P. Cao, W. Li, X. Zhang, X. Wei, Synthesis and magnetic properties of size-controlled FeNi alloy nanoparticles attached on multiwalled carbon nanotubes, *Journal of Physics and Chemistry of Solids.* 71 (2010) 290–295. https://doi.org/10.1016/j.jpcs.2009.12.079.

[89] White, C. W., S. P. Withrow, K. D. Sorge, A. Meldrum, J. D. Budai, J. R. Thompson, L. A. Boatner, Oriented ferromagnetic Fe-Pt alloy nanoparticles produced in $Al_2O_3$ by ion-beam synthesis, *J Appl Phys.* 93 (2003) 5656–5669. https://doi.org/10.1063/1.1565691.

[90] Faraji, M., Y. Yamini, M. Rezaee, Magnetic nanoparticles: synthesis, stabilization, functionalization, characterization, and applications, *Journal of the Iranian Chemical Society.* 7 (2010) 1–37. https://doi.org/10.1007/BF03245856.

[91] Indira, T. K., P. K. Lakshmi, Magnetic nanoparticles–a review, *Int. J. Pharm. Sci. Nanotechnol.* 3 (2010) 1035–1042. https://doi.org/10.37285/ijpsn.2010.3.3.1.

[92] Ahn, T., J. H. Kim, H. M. Yang, J. W. Lee, J. D. Kim, Formation pathways of magnetite nanoparticles by coprecipitation method, *The Journal of Physical Chemistry C.* 116 (2012) 6069–6076. https://doi.org/10.1021/jp211843g.

[93] Bee, A., R. Massart, S. Neveu, Synthesis of very fine maghemite particles, *J Magn Magn Mater.* 149 (1995) 6–9. https://doi.org/10.1016/0304-8853(95)00317-7.

[94] Petcharoen, K., A. Sirivat, Synthesis and characterization of magnetite nanoparticles via the chemical co-precipitation method, *Materials Science and Engineering: B.* 177 (2012) 421–427. https://doi.org/10.1016/j.mseb.2012.01.003.

[95] Peternele, W. S., V. M. Fuentes, M. L. Fascineli, J. R. da Silva, R. C. Silva, C.M. Lucci, R. B. de Azevedo, Experimental investigation of the coprecipitation method:

an approach to obtain magnetite and maghemite nanoparticles with improved properties, *J Nanomater.* 2014 (2014) 94. https://doi.org/10.1155/2014/682985.

[96] Roth, H. C., S. P. Schwaminger, M. Schindler, F. E. Wagner, S. Berensmeier, Influencing factors in the CO-precipitation process of superparamagnetic iron oxide nano particles: a model based study, *J Magn Magn Mater.* 377 (2015) 81–89. https://doi.org/10.1016/j.jmmm.2014.10.074.

[97] Iida, H., K. Takayanagi, T. Nakanishi, T. Osaka, Synthesis of $Fe_3O_4$ nanoparticles with various sizes and magnetic properties by controlled hydrolysis, *J Colloid Interface Sci.* 314 (2007) 274–280. https://doi.org/10.1016/j.jcis.2007.05.047.

[98] Baumgartner, J., A. Dey, P. H. H. Bomans, C. Le Coadou, P. Fratzl, N. A. J. M. Sommerdijk, D. Faivre, Nucleation and growth of magnetite from solution, *Nat Mater.* 12 (2013) 310. https://doi.org/10.1038/nmat3558.

[99] Willis, A. L., N. J. Turro, S. O'Brien, Spectroscopic characterization of the surface of iron oxide nanocrystals, *Chemistry of Materials.* 17 (2005) 5970–5975. https://doi.org/10.1021/cm051370v.

[100] Qiu, X. Synthesis and characterization of magnetic nano particles, *Chin J Chem.* 18 (2000) 834–837. https://doi.org/10.1002/cjoc.20000180607.

[101] Murray, Cb., D. J. Norris, M. G. Bawendi, Synthesis and characterization of nearly monodisperse CdE (E = sulfur, selenium, tellurium) semiconductor nanocrystallites, *J Am Chem Soc.* 115 (1993) 8706–8715. https://doi.org/10.1021/ja00072a025.

[102] Peng, X., J. Wickham, A.P. Alivisatos, Kinetics of II-VI and III-V colloidal semiconductor nanocrystal growth:"focusing" of size distributions, *J Am Chem Soc.* 120 (1998) 5343–5344. https://doi.org/10.1021/ja9805425.

[103] O'Brien, S., L. Brus, C. B. Murray, Synthesis of monodisperse nanoparticles of barium titanate: toward a generalized strategy of oxide nanoparticle synthesis, *J Am Chem Soc.* 123 (2001) 12085–12086. https://doi.org/10.1021/ja011414a.

[104] Lu, A. H., E. L. Salabas, F. Schüth, Magnetic nanoparticles: Synthesis, protection, functionalization, and application, *Angewandte Chemie - International Edition.* 46 (2007) 1222–1244. https://doi.org/10.1002/anie.200602866.

[105] Pankhurst, Q. A., J. Connolly, S.K. Jones, J. Dobson, *Applications of magnetic nanoparticles in biomedicine*, 36 (2003). https://doi.org/10.1088/0022-3727/36/13/201.

[106] Chen, Y., D. L. Peng, D. Lin, X. Luo, Preparation and magnetic properties of nickel nanoparticles via the thermal decomposition of nickel organometallic precursor in alkylamines, *Nanotechnology.* 18 (2007) 505703. https://doi.org/10.1088/0957-4484/18/50/505703.

[107] Jana, N. R., Y. Chen, X. Peng, Size-and shape-controlled magnetic (Cr, Mn, Fe, Co, Ni) oxide nanocrystals via a simple and general approach, *Chemistry of Materials.* 16 (2004) 3931–3935. https://doi.org/10.1021/cm049221k.

[108] Park, J., K. An, Y. Hwang, J. G. Park, H. J. Noh, J. Y. Kim, J. H. Park, N. M. Hwang, T. Hyeon, Ultra-large-scale syntheses of monodisperse nanocrystals, *Nat Mater.* 3 (2004) 891. https://doi.org/10.1038/nmat1251.

[109] Li, Z., Q. Sun, M. Gao, Preparation of water-soluble magnetite nanocrystals from hydrated ferric salts in 2-pyrrolidone: mechanism leading to $Fe_3O_4$, *Angewandte*

*Chemie International Edition.* 44 (2005) 123–126. https://doi.org/10.1002/anie. 200460715.

[110] Butter, K., K. Kassapidou, G. J. Vroege, A. P. Philipse, Preparation and properties of colloidal iron dispersions, *J Colloid Interface Sci.* 287 (2005) 485–495. https://doi.org/10.1016/j.jcis.2005.02.014.

[111] Mao, B., Z. Kang, E. Wang, S. Lian, L. Gao, C. Tian, C. Wang, Synthesis of magnetite octahedrons from iron powders through a mild hydrothermal method, *Mater Res Bull.* 41 (2006) 2226–2231. https://doi.org/10.1016/j.materresbull.2006.04.037.

[112] Giri, S., S. Samanta, S. Maji, S. Ganguli, A. Bhaumik, Magnetic properties of α-$Fe_2O_3$ nanoparticle synthesized by a new hydrothermal method, *J Magn Magn Mater.* 285 (2005) 296–302. https://doi.org/10.1016/j.jmmm.2004.08.007.

[113] Wang, X., J. Zhuang, Q. Peng, Y. Li, A general strategy for nanocrystal synthesis, *Nature.* 437 (2005) 121. https://doi.org/10.1038/nature03968.

[114] Wang, J., J. Sun, Q. Sun, Q. Chen, One-step hydrothermal process to prepare highly crystalline Fe3O4 nanoparticles with improved magnetic properties, *Mater Res Bull.* 38 (2003) 1113–1118. https://doi.org/10.1016/S0025-5408(03)00129-6.

[115] Wang, J., F. Ren, R. Yi, A. Yan, G. Qiu, X. Liu, Solvothermal synthesis and magnetic properties of size-controlled nickel ferrite nanoparticles, *J Alloys Compd.* 479 (2009) 791–796. https:// doi.org/10.1016/j.jallcom.2009.01.059.

[116] Xu, C., A. S. Teja, Continuous hydrothermal synthesis of iron oxide and PVA-protected iron oxide nanoparticles, *J Supercrit Fluids.* 44 (2008) 85–91. https://doi.org/10.1016/j.supflu.2007.09.033.

[117] Komarneni, S., H. Katsuki, Nanophase materials by a novel microwave-hydrothermal process, *Pure and Applied Chemistry.* 74 (2002) 1537–1543. https://doi.org/10.1351/pac200274091537.

[118] Sreeja, V., P. A. Joy, Microwave–hydrothermal synthesis of γ-$Fe_2O_3$ nanoparticles and their magnetic properties, *Mater Res Bull.* 42 (2007) 1570–1576. https://doi.org/10.1016/j.materresbull.2006.11.014.

[119] Lopez Perez, J. A., M. A. Lopez Quintela, J. Mira, J. Rivas, S. W. Charles, Advances in the preparation of magnetic nanoparticles by the microemulsion method, *J Phys Chem B.* 101 (1997) 8045–8047. https://doi.org/10.1021/jp972046t.

[120] Dresco, P. A., V. S. Zaitsev, R. J. Gambino, B. Chu, Preparation and properties of magnetite and polymer magnetite nanoparticles, *Langmuir.* 15 (1999) 1945–1951. https://doi.org/10.1021/la980971g.

[121] Santra, S., R. Tapec, N. Theodoropoulou, J. Dobson, A. Hebard, W. Tan, Synthesis and characterization of silica-coated iron oxide nanoparticles in microemulsion: the effect of nonionic surfactants, *Langmuir.* 17 (2001) 2900–2906. https://doi.org/10.1021/la0008636.

[122] Liu, C., B. Zou, A. J. Rondinone, Z. J. Zhang, Reverse micelle synthesis and characterization of superparamagnetic $MnFe_2O_4$ spinel ferrite nanocrystallites, *J Phys Chem B.* 104 (2000) 1141–1145. https://doi.org/10.1021/jp993552g.

# References

[123] Gupta, A. K., M. Gupta, Synthesis and surface engineering of iron oxide nanoparticles for biomedical applications, *Biomaterials.* 26 (2005) 3995–4021. https://doi.org/10.1016/j.biomaterials.2004.10.012.

[124] Abou-Hassan, A., R. Bazzi, V. Cabuil, Multistep continuous-flow microsynthesis of magnetic and fluorescent γ-$Fe_2O_3$@$SiO_2$ core/shell nanoparticles, *Angewandte Chemie International Edition.* 48 (2009) 7180–7183. https://doi.org/10.1002/anie.200902181.

[125] Lemine, O. M., K. Omri, B. Zhang, L. El Mir, M. Sajieddine, A. Alyamani, M. Bououdina, Sol–gel synthesis of 8 nm magnetite ($Fe_3O_4$) nanoparticles and their magnetic properties, *Superlattices Microstruct.* 52 (2012) 793–799. https://doi.org/10.1016/j.spmi.2012.07.009.

[126] Qi, H., B. Yan, W. Lu, C. Li, Y. Yang, A non-alkoxide sol-gel method for the preparation of magnetite ($Fe_3O_4$) nanoparticles, *Curr Nanosci.* 7 (2011) 381–388. https://doi.org/10.2174/157341311795542426.

[127] Woo, K., H. J. Lee, J. Ahn, Y. S. Park, Sol–gel mediated synthesis of $Fe_2O_3$ nanorods, *Advanced Materials.* 15 (2003) 1761–1764. https://doi.org/10.1002/adma.200305561.

[128] Landau, L., E. Lifshitz, On the theory of the dispersion of magnetic permeability in ferromagnetic bodies, in: *Perspectives in Theoretical Physics*, Elsevier, 1992: pp. 51–65. https://doi.org/10.1016/B978-0-08-036364-6.50008-9.

[129] Dutz, S. Are magnetic multicore nanoparticles promising candidates for biomedical applications?, *IEEE Trans Magn.* 52 (2016) 1–3. https://doi.org/10.1109/TMAG.2016.2570745.

[130] Butler, R. F., S. K. Banerjee, Theoretical single-domain grain size range in magnetite and titanomagnetite, *J Geophys Res.* 80 (1975) 4049–4058. https://doi.org/10.1029/JB080i029p04049.

[131] Lartigue, L., P. Hugounenq, D. Alloyeau, S. P. Clarke, M. Lévy, J. C. Bacri, R. Bazzi, D. F. Brougham, C. Wilhelm, F. Gazeau, Cooperative organization in iron oxide multi-core nanoparticles potentiates their efficiency as heating mediators and MRI contrast agents, *ACS Nano.* 6 (2012) 10935–10949. https://doi.org/10.1021/nn304477s.

[132] Lopez, J. A., F. González, F. A. Bonilla, G. Zambrano, M. E. Gómez, Synthesis and characterization of $Fe_3O_4$ magnetic nanofluid, *Revista Latinoamericana de Metalurgia y Materiales.* 30 (2010) 60–66.

[133] Stephens, J. R., J. S. Beveridge, M. E. Williams, Analytical methods for separating and isolating magnetic nanoparticles, *Physical Chemistry Chemical Physics.* 14 (2012) 3280–3289. https://doi.org/10.1039/C2CP22982J.

[134] Zhao, W., R. Cheng, J. R. Miller, L. Mao, Label-Free Microfluidic Manipulation of Particles and Cells in Magnetic Liquids, *Adv Funct Mater.* 26 (2016) 3916–3932. https://doi.org/10.1002/adfm.201504178.

[135] Hejazian, M., N. T. Nguyen, Magnetofluidic concentration and separation of non-magnetic particles using two magnet arrays, *Biomicrofluidics.* 10 (2016) 44103. https://doi.org/10.1063/1.4955421.

[136] Wu, W., X. H. Xiao, S. F. Zhang, T. C. Peng, J. Zhou, F. Ren, C. Z. Jiang, Synthesis and Magnetic Properties of Maghemite (gamma-$Fe_{(2)}O_{(3)}$)) *Short-Nanotubes,*

*Nanoscale Res Lett.* 5 (2010) 1474–1479. https://doi.org/10.1007/s11671-010-9664-4.

[137] Wu, W., X. Xiao, S. Zhang, F. Ren, C. Jiang, Facile method to synthesize magnetic iron oxides/TiO2 hybrid nanoparticles and their photodegradation application of methylene blue, *Nanoscale Res Lett.* 6 (2011) 533. https://doi.org/10.1186/1556-276X-6-533.

[138] Xuan, S., Y. X. J. Wang, J. C. Yu, K. Cham-Fai Leung, Tuning the Grain Size and Particle Size of Superparamagnetic $Fe_3O_4$ Microparticles, *Chemistry of Materials.* 21 (2009) 5079–5087. https://doi.org/10.1021/cm901618m.

[139] Demortière, A., P. Panissod, B. P. Pichon, G. Pourroy, D. Guillon, B. Donnio, S. Bégin-Colin, Size-dependent properties of magnetic iron oxide nanocrystals, *Nanoscale.* 3 (2011) 225–232. https://doi.org/10.1039/C0NR00521E.

[140] Santoyo Salazar, J., L. Perez, O. de Abril, L. Truong Phuoc, D. Ihiawakrim, M. Vazquez, J. M. Greneche, S. Begin-Colin, G. Pourroy, Magnetic Iron Oxide Nanoparticles in 10−40 nm Range: Composition in Terms of Magnetite/Maghemite Ratio and Effect on the Magnetic Properties, *Chemistry of Materials.* 23 (2011) 1379–1386. https://doi.org/10.1021/cm103188a.

[141] Yun, H., X. Liu, T. Paik, D. Palanisamy, J. Kim, W. D. Vogel, A. J. Viescas, J. Chen, G. C. Papaefthymiou, J. M. Kikkawa, M. G. Allen, C. B. Murray, Size- and Composition-Dependent Radio Frequency Magnetic Permeability of Iron Oxide Nanocrystals, *ACS Nano.* 8 (2014) 12323–12337. https://doi.org/10.1021/nn504711g.

[142] Guardia, P., A. Riedinger, S. Nitti, G. Pugliese, S. Marras, A. Genovese, M. E. Materia, C. Lefevre, L. Manna, T. Pellegrino, One pot synthesis of monodisperse water soluble iron oxide nanocrystals with high values of the specific absorption rate, *J Mater Chem B.* 2 (2014) 4426–4434. https://doi.org/10.1039/C4TB00061G.

[143] Liu, L., H. Z. Kou, W. Mo, H. Liu, Y. Wang, Surfactant-Assisted Synthesis of α-$Fe_2O_3$ Nanotubes and Nanorods with Shape-Dependent Magnetic Properties, *J Phys Chem B.* 110 (2006) 15218–15223. https://doi.org/10.1021/jp0627473.

[144] Wu, W., S. Yang, J. Pan, L. Sun, J. Zhou, Z. Dai, X. Xiao, H. Zhang, C. Jiang, Metal ion-mediated synthesis and shape-dependent magnetic properties of single-crystalline α-$Fe_2O_3$ nanoparticles, *Cryst Eng Comm.* 16 (2014) 5566–5572. https://doi.org/10.1039/C4CE00048J.

[145] Song, M., Y. Zhang, S. Hu, L. Song, J. Dong, Z. Chen, N. Gu, Influence of morphology and surface exchange reaction on magnetic properties of monodisperse magnetite nanoparticles, *Colloids Surf A Physicochem Eng Asp.* 408 (2012) 114–121. https://doi.org/https://doi.org/10.1016/j.colsurfa.2012.05.039.

[146] Albanese, A., P. S. Tang, W. C. W. Chan, The effect of nanoparticle size, shape, and surface chemistry on biological systems, *Annu Rev Biomed Eng.* 14 (2012) 1–16. https://doi.org/10.1146/annurev-bioeng-071811-150124.

[147] Golas, P. L., S. Louie, G. V Lowry, K. Matyjaszewski, R. D. Tilton, Comparative Study of Polymeric Stabilizers for Magnetite Nanoparticles Using ATRP, *Langmuir.* 26 (2010) 16890–16900. https://doi.org/10.1021/la103098q.

# References

[148] Phenrat, T., N. Saleh, K. Sirk, R. D. Tilton, G. V Lowry, Aggregation and Sedimentation of Aqueous Nanoscale Zerovalent Iron Dispersions, *Environ Sci Technol.* 41 (2007) 284–290. https://doi.org/10.1021/es061349a.

[149] Henry, C., J. P. Minier, J. Pozorski, G. Lefèvre, A New Stochastic Approach for the Simulation of Agglomeration between Colloidal Particles, *Langmuir.* 29 (2013) 13694–13707. https://doi.org/10.1021/la403615w.

[150] Yeap, S. P., J. Lim, B. S. Ooi, A. L. Ahmad, Agglomeration, colloidal stability, and magnetic separation of magnetic nanoparticles: collective influences on environmental engineering applications, *Journal of Nanoparticle Research.* 19 (2017) 368. https://doi.org/10.1007/s11051-017-4065-6.

[151] Dickson, D., G. Liu, C. Li, G. Tachiev, Y. Cai, Dispersion and stability of bare hematite nanoparticles: Effect of dispersion tools, nanoparticle concentration, humic acid and ionic strength, *Science of The Total Environment.* 419 (2012) 170–177. https://doi.org/https://doi.org/10.1016/j.scitotenv.2012.01.012.

[152] Wu, W., G. Ichihara, Y. Suzuki, K. Izuoka, S. Oikawa-Tada, J. Chang, K. Sakai, K. Miyazawa, D. Porter, V. Castranova, M. Kawaguchi, S. Ichihara, Dispersion method for safety research on manufactured nanomaterials, *Ind Health.* 52 (2014) 54–65. https://doi.org/10.2486/indhealth.2012-0218.

[153] Mikhaylova, M., Y. S. Jo, D. K. Kim, N. Bobrysheva, Y. Andersson, T. Eriksson, M. Osmolowsky, V. Semenov, M. Muhammed, The Effect of Biocompatible Coating Layers on Magnetic Properties of Superparamagnetic Iron Oxide Nanoparticles, *Hyperfine Interact.* 156 (2004) 257–263. https://doi.org/10.1023/B:HYPE.0000043238.36641.7b.

[154] Xiong, Z., D. Zhao, G. Pan, Rapid and complete destruction of perchlorate in water and ion-exchange brine using stabilized zero-valent iron nanoparticles, *Water Res.* 41 (2007) 3497–3505. https://doi.org/https://doi.org/10.1016/j.watres.2007.05.049.

[155] Hakami, O., Y. Zhang, C. J. Banks, Thiol-functionalised mesoporous silica-coated magnetite nanoparticles for high efficiency removal and recovery of Hg from water, *Water Res.* 46 (2012) 3913–3922. https://doi.org/https://doi.org/10.1016/j.watres.2012.04.032.

[156] Sirk, K. M., N. B. Saleh, T. Phenrat, H. J. Kim, B. Dufour, J. Ok, P. L. Golas, K. Matyjaszewski, G. V Lowry, R. D. Tilton, Effect of Adsorbed Polyelectrolytes on Nanoscale Zero Valent Iron Particle Attachment to Soil Surface Models, *Environ Sci Technol.* 43 (2009) 3803–3808. https://doi.org/10.1021/es803589t.

[157] Yantasee, W., C. L. Warner, T. Sangvanich, R. S. Addleman, T. G. Carter, R. J. Wiacek, G. E. Fryxell, C. Timchalk, M. G. Warner, Removal of Heavy Metals from Aqueous Systems with Thiol Functionalized Superparamagnetic Nanoparticles, *Environ Sci Technol.* 41 (2007) 5114–5119. https://doi.org/10.1021/es0705238.

[158] Chang, Y. C., D. H. Chen, Preparation and adsorption properties of monodisperse chitosan-bound $Fe_3O_4$ magnetic nanoparticles for removal of Cu(II) ions, *J Colloid Interface Sci.* 283 (2005) 446–451. https://doi.org/https://doi.org/10.1016/j.jcis.2004.09.010.

# References

[159] Geng, B., Z. Jin, T. Li, X. Qi, Kinetics of hexavalent chromium removal from water by chitosan-Fe$^0$ nanoparticles, *Chemosphere*. 75 (2009) 825–830. https://doi.org/ https://doi.org/10.1016/j.chemosphere.2009.01.009.

[160] Wang, J., S. Zheng, Y. Shao, J. Liu, Z. Xu, D. Zhu, Amino-functionalized Fe$_3$O$_4$@SiO$_2$ core–shell magnetic nanomaterial as a novel adsorbent for aqueous heavy metals removal, *J Colloid Interface Sci*. 349 (2010) 293–299. https://doi.org/ https://doi.org/10.1016/j.jcis.2010.05.010.

[161] Wang, Q., H. Qian, Y. Yang, Z. Zhang, C. Naman, X. Xu, Reduction of hexavalent chromium by carboxymethyl cellulose-stabilized zero-valent iron nanoparticles, *J Contam Hydrol*. 114 (2010) 35–42. https://doi.org/https://doi.org/10.1016/ j.jconhyd.2010.02.006.

[162] Chen, D. H., S. H. Huang, Fast separation of bromelain by polyacrylic acid-bound iron oxide magnetic nanoparticles, *Process Biochemistry*. 39 (2004) 2207–2211. https://doi.org/https://doi.org/10.1016/j.procbio.2003.11.014.

[163] Huang, S. H., M. H. Liao, D. H. Chen, Fast and efficient recovery of lipase by polyacrylic acid-coated magnetic nano-adsorbent with high activity retention, Sep *Purif Technol*. 51 (2006) 113–117. https://doi.org/https://doi.org/10.1016/ j.seppur.2006.01.003.

[164] Liao, M., K. Wu, D. Chen, Fast adsorption of crystal violet on polyacrylic acid-bound magnetic nanoparticles, *Sep Sci Technol*. 39 (2005) 1563–1575. https://doi.org/10.1081/SS-120030802.

[165] Mahdavian, A. R., M. A. S. Mirrahimi, Efficient separation of heavy metal cations by anchoring polyacrylic acid on superparamagnetic magnetite nanoparticles through surface modification, *Chemical Engineering Journal*. 159 (2010) 264–271. https://doi.org/https://doi.org/10.1016/j.cej.2010.02.041.

[166] F. He, D. Zhao, C. Paul, Field assessment of carboxymethyl cellulose stabilized iron nanoparticles for *in situ* destruction of chlorinated solvents in source zones, *Water Res*. 44 (2010) 2360–2370. https://doi.org/https://doi.org/10.1016/j.watres. 2009.12.041.

[167] Phenrat, T., N. Saleh, K. Sirk, H. J. Kim, R. D. Tilton, G. V Lowry, Stabilization of aqueous nanoscale zerovalent iron dispersions by anionic polyelectrolytes: adsorbed anionic polyelectrolyte layer properties and their effect on aggregation and sedimentation, *Journal of Nanoparticle Research*. 10 (2008) 795–814. https://doi.org/10.1007/s11051-007-9315-6.

[168] Tang, S. C. N., I. M. C. Lo, Magnetic nanoparticles: Essential factors for sustainable environmental applications, *Water Res*. 47 (2013) 2613–2632. https://doi.org/https://doi.org/10.1016/j.watres.2013.02.039.

[169] Liu, J., Z. Zhao, G. Jiang, Coating Fe$_3$O$_4$ Magnetic Nanoparticles with Humic Acid for High Efficient Removal of Heavy Metals in Water, *Environ Sci Technol*. 42 (2008) 6949–6954. https://doi.org/10.1021/es800924c.

[170] Sakulchaicharoen, N., D. M. O'Carroll, J. E. Herrera, Enhanced stability and dechlorination activity of pre-synthesis stabilized nanoscale FePd particles, *J Contam Hydrol*. 118 (2010) 117–127. https://doi.org/https://doi.org/10.1016/ j.jconhyd.2010.09.004.

## References

[171] Ali, A., M. Z. Hira Zafar, I. ul Haq, A. R. Phull, J. S. Ali, A. Hussain, Synthesis, characterization, applications, and challenges of iron oxide nanoparticles, *Nanotechnol Sci Appl*. 9 (2016) 49. https://doi.org/10.2147/NSA.S99986.

[172] Laurent, S., D. Forge, M. Port, A. Roch, C. Robic, L. Vander Elst, R. N. Muller, Magnetic Iron Oxide Nanoparticles: Synthesis, Stabilization, Vectorization, Physicochemical Characterizations, and Biological Applications, *Chem Rev*. 108 (2008) 2064–2110. https://doi.org/10.1021/cr068445e.

[173] Shiba, K., T. Sugiyama, T. Takei, G. Yoshikawa, Controlled growth of silica–titania hybrid functional nanoparticles through a multistep microfluidic approach, *Chemical Communications*. 51 (2015) 15854–15857. https://doi.org/10.1039/C5CC07230A.

[174] Simmons, M. D., N. Jones, D. J. Evans, C. Wiles, P. Watts, S. Salamon, M. E. Castillo, H. Wende, D. C. Lupascu, M. G. Francesconi, Doping of inorganic materials in microreactors–preparation of Zn doped $Fe_3O_4$ nanoparticles, *Lab Chip*. 15 (2015) 3154–3162. https://doi.org/10.1039/C5LC00287G.

[175] Bhattacharjee, A., S. Gumma, M. K. Purkait, $Fe_3O_4$ promoted metal organic framework MIL-100 (Fe) for the controlled release of doxorubicin hydrochloride, *Microporous and Mesoporous Materials*. 259 (2018) 203–210. https://doi.org/10.1016/j.micromeso.2017.10.020.

[176] El Ghandoor, H., H. M. Zidan, M. M. H. Khalil, M. I. M. Ismail, Synthesis and some physical properties of magnetite ($Fe_3O_4$) nanoparticles, *Int. J. Electrochem. Sci*. 7 (2012) 5734–5745.

[177] Hong, R. Y., J. H. Li, H. Z. Li, J. Ding, Y. Zheng, D. G. Wei, Synthesis of Fe3O4 nanoparticles without inert gas protection used as precursors of magnetic fluids, *J Magn Magn Mater*. 320 (2008) 1605–1614. https://doi.org/10.1016/j.jmmm.2008.01.015.

[178] Wei, Y., B. Han, X. Hu, Y. Lin, X. Wang, X. Deng, Synthesis of Fe3O4 nanoparticles and their magnetic properties, *Procedia Eng*. 27 (2012) 632–637. https://doi.org/10.1016/j.proeng.2011.12.498.

[179] Khalil, M. I. Co-precipitation in aqueous solution synthesis of magnetite nanoparticles using iron (III) salts as precursors, *Arabian Journal of Chemistry*. 8 (2015) 279–284. https://doi.org/10.1016/j.arabjc.2015.02.008.

[180] Loekitowati Hariani, P., M. Faizal, R. Ridwan, M. Marsi, D. Setiabudidaya, Synthesis and properties of Fe3O4 nanoparticles by co-precipitation method to removal procion dye, *International Journal of Environmental Science and Development*. 4 (2013) 336–340. https://doi.org/10.7763/IJESD.2013.V4.366.

[181] Kim, D. K., Y. Zhang, W. Voit, K. V Rao, M. Muhammed, Synthesis and characterization of surfactant-coated superparamagnetic monodispersed iron oxide nanoparticles, *J Magn Magn Mater*. 225 (2001) 30–36. https://doi.org/https://doi.org/10.1016/S0304-8853(00)01224-5.

[182] Agnihotri, P., Lad, V.N. Magnetic nanofluid: synthesis and characterization. *Chem. Pap*. 74, 3089–3100 (2020). https://doi.org/10.1007/s11696-020-01138-w

[183] Sun, J., S. Zhou, P. Hou, Y. Yang, J. Weng, X. Li, M. Li, *Synthesis and characterization of biocompatible $Fe_3O_4$ nanoparticles*, 2007. https://doi.org/10.1002/jbm.a.30909.

# References

[184] Shen, Y. F., J. Tang, Z. H. Nie, Y. D. Wang, Y. Ren, L. Zuo, Preparation and application of magnetic $Fe_3O_4$ nanoparticles for wastewater purification, *Sep Purif Technol.* 68 (2009) 312–319. https://doi.org/10.1016/j.seppur.2009.05.020.

[185] Sun, Y. P., X. Li, J. Cao, W. Zhang, H. P. Wang, Characterization of zero-valent iron nanoparticles, *Adv Colloid Interface Sci.* 120 (2006) 47–56. https://doi.org/10.1016/j.cis.2006.03.001.

[186] Pislaru-Danescu, L., A. Morega, G. Telipan, V. Stoica, Nanoparticles of ferrofluid $Fe_3O_4$ synthetised by coprecipitation method used in microactuation process, *Optoelectronics and Advanced Materials, Rapid Communications.* 4 (2010) 1182–1186.

[187] Maaz, K., S. Karim, A. Mumtaz, S. K. Hasanain, J. Liu, J. L. Duan, Synthesis and magnetic characterization of nickel ferrite nanoparticles prepared by co-precipitation route, *J Magn Magn Mater.* 321 (2009) 1838–1842. https://doi.org/10.1016/j.jmmm.2008.11.098.

[188] Shete, P. B., R. M. Patil, B. M. Tiwale, S. H. Pawar, Water dispersible oleic acid-coated Fe3O4 nanoparticles for biomedical applications, *J Magn Magn Mater.* 377 (2015) 406–410. https://doi.org/https://doi.org/10.1016/j.jmmm.2014.10.137.

[189] Tao, K., H. Dou, K. Sun, Interfacial coprecipitation to prepare magnetite nanoparticles: Concentration and temperature dependence, *Colloids Surf A Physicochem Eng Asp.* 320 (2008) 115–122. https://doi.org/10.1016/j.colsurfa.2008.01.051.

[190] Hu, C., Z. Gao, X. Yang, One-pot low temperature synthesis of $MFe_2O_4$ (M= Co, Ni, Zn) superparamagnetic nanocrystals, *J Magn Magn Mater.* 320 (2008) L70–L73. https://doi.org/10.1016/j.jmmm.2007.12.006.

[191] Bini, R. A., R. F. C. Marques, F. J. Santos, J. A. Chaker, M. Jafelicci Jr, Synthesis and functionalization of magnetite nanoparticles with different amino-functional alkoxysilanes, *J Magn Magn Mater.* 324 (2012) 534–539. https://doi.org/10.1016/j.jmmm.2011.08.035.

[192] Mahdavi, M., M. Ahmad, M. Haron, F. Namvar, B. Nadi, M. Rahman, J. Amin, Synthesis, surface modification and characterisation of biocompatible magnetic iron oxide nanoparticles for biomedical applications, *Molecules.* 18 (2013) 7533–7548. https://doi.org/10.3390/molecules18077533.

[193] Deshmukh, A. R., A. Gupta, B. S. Kim, Ultrasound Assisted Green Synthesis of Silver and Iron Oxide Nanoparticles Using Fenugreek Seed Extract and Their Enhanced Antibacterial and Antioxidant Activities, *Biomed Res Int.* 2019 (2019). https://doi.org/10.1155/2019/1714358.

[194] Sulistyaningsih, T., S. J. Santosa, D. Siswanta, B. Rusdiarso, *Synthesis and Characterization of Magnetites Obtained from Mechanically and Sonochemically Assissted Co-precipitation and Reverse Co-precipitation Methods*, 5 (2017) 3–6. https://doi.org/10.18178/ijmmm.2017.5.1.280.

[195] Sarkar, Z.K., F. K. Sarkar, Synthesis and Magnetic Properties Investigations of $Fe_3O_4$ *Int. J. Nanosci. Nanotechnol.* 7 (2011) 197–200.

[196] Wu, W., X. Xiao, S. Zhang, J. Zhou, L. Fan, F. Ren, C. Jiang, Large-scale and controlled synthesis of iron oxide magnetic short nanotubes: shape evolution,

growth mechanism, and magnetic properties, *The Journal of Physical Chemistry C.* 114 (2010) 16092–16103. https://doi.org/10.1021/jp1010154.

[197] Drbohlavova, J., R. Hrdy, V. Adam, R. Kizek, O. Schneeweiss, J. Hubalek, C. Republic, C. Republic, C. Republic, *Preparation and Properties of Various Magnetic Nanoparticles*, (2009) 2352–2362. https://doi.org/10.3390/s90402352.

[198] Lassoued, A., M. S. Lassoued, B. Dkhil, S. Ammar, A. Gadri, Synthesis, photoluminescence and Magnetic properties of iron oxide ($\alpha$-Fe2O3) nanoparticles through precipitation or hydrothermal methods, *Physica E Low Dimens Syst Nanostruct.* 101 (2018) 212–219. https://doi.org/10.1016/j.physe.2018.04.009.

[199] Pianciola, B. N. E. Lima Jr, H. E. Troiani, L. C. C. M. Nagamine, R. Cohen, R. D. Zysler, Size and surface effects in the magnetic order of CoFe2O4 nanoparticles, *J Magn Magn Mater.* 377 (2015) 44–51. https://doi.org/10.1016/j.jmmm.2014.10.054.

[200] Neyaz, N., W. A. Siddiqui, K. K. Nair, Application of surface functionalized iron oxide nanomaterials as a nanosorbents in extraction of toxic heavy metals from ground water: a review, *Int J Environ Sci.* 4 (2014) 472. https://doi.org/10.6088/ijes.2014040400004.

[201] Ma, K. Q., Liu, J., Nano liquid-metal fluid as ultimate coolant, *Phys Lett A*, 361(3) (2007) 252–256. https://doi.org/10.1016/J.PHYSLETA.2006.09.041.

[202] Bairwa, D. K., K. K. Upman, G. Kantak, Nanofluids and its applications, *Int J Eng Manag Sci* 2 (2015) 14–17.

[203] Stoian, F. D., S. Holotescu, Experimental study of cooling enhancement using a Fe3O4 magnetic nanofluid, in an applied magnetic field, *J Phys Conf Series*, 547 (1) (2014). https://doi.org/10.1088/1742-6596/547/1/012044.

[204] Patra, K., A. Sengupta, V. K. Mittal, T. P. Valsala, Emerging functionalized magnetic nanoparticles: from synthesis to nuclear fuel cycle application: Where do we stand after two decades? *Mater Today Sus* (2023) 100489. https://doi.org/10.1016/j.mtsust.2023.100489.

[205] Lakshmanan, R., C. Okoli, M. Boutonnet, S. Jaras, G. K. Rajarao, Microemulsion prepared magnetic nanoparticles for phosphate removal: Time efficient studies, *J Environ Chem Eng* 2 (1) (2014) 185-189. https://doi.org/10.1016/j.jece.2013.12.008.

[206] Ming, Z., L. Zhongliang, M. Guoyuan, C. Shuiyuan, The experimental study on flat plate heat pipe of magnetic working fluid, *Exp Therm Fluid Sci* 33(7) (2009) 1100–1105. https://doi.org/https://doi.org/10.1016/j.expthermflusci.2009.06.009.

[207] Bahiraei, M., M. Hangi, Flow and heat transfer characteristics of magnetic nanofluids: A review, *Magn Mater* 374 (2015) 125-138. https://doi.org/10.1016/j.jmmm.2014.08.004.

[208] Chiang, Y. C., J. J. Chieh, C. C. Ho, The magnetic-nanofluid heat pipe with superior thermal properties through magnetic enhancement, *Nanoscale Res Lett* 7(1) (2012) 322. https://doi.org/10.1186/1556-276X-7-322.

[209] Taslimifar, M., M. Mohammadi, H. Afshin, M. H. Saidi, M. B. Shafii, Overall thermal performance of ferrofluidic open loop pulsating heat pipes: An experimental approach, *Int J Therm Sci* 65 (2013) 234–241. https://doi.org/10.1016/j.ijthermalsci.2012.10.016.

[210] Fumoto, K., H. Yamagishi, M. Ikegawa, A mini heat transport device based on thermo-sensitive magnetic fluid, *Nanoscale Microscale Thermophys Eng* 11(1–2) (2007) 201–210. https://doi.org/10.1080/15567260701333869.

[211] Philip, J., P. D. Shima, B. Raj, Nanofluid with tunable thermal properties, *Appl Phys Lett* 92(4) (2008) 1–4. https://doi.org/10.1063/1.2838304.

[212] Agnihotri, P., R. Tala, P. Doot, V. N. Lad, Microfuidics for selective concentration of nanofuid streams containing magnetic nanoparticles, *Sep Sci Technol* 54 (2019) 289–292. https://doi.org/10.1080/01496395.2018.1529041.

[213] Agnihotri, P., V. N. Lad, Controlled release and separation of magnetic nanoparticles using microfuidics by varying bifurcation angle of microchannels, *J Inorg Organomet Polym Mater* 29 (2019) 309–315. https://doi.org/10.1007/s10904-018-1000-y.

[214] Agnihotri, P., V. N. Lad, Controlling the interface in microfuidic fow by varying orientations of microchannels with reference to the magnetic field, *J Braz Soc Mech Sci Eng* 45 (2023) 272. https://doi.org/10.1007/s40430-023-04167-0.

[215] Rosensweig, R. E. Magnetic Fluids, *Ann Rev Fluid Mechanics* 19(1) (1987) 437–461. https://doi.org/10.1146/annurev.fl.19.010187.002253.

[216] Kim, Y. S., K. Nakatsuka, T. Fujita, T. Atarashi, Application of hydrophilic magnetic fluid to oil seal, *J Magn Magn Mater* 201 (1–3) (1999) 361–363. https://doi.org/10.1016/S0304-8853(99)00117-1.

[217] Tong, S., H. Zhu, G. Bao, Magnetic iron oxide nanoparticles for disease detection and therapy, *Mater Today* 31 (2019) 86–99. https://doi.org/10.1016/j.mattod.2019.06.003.

[218] Mody V.V., A. Singh, B. Wesley, Basics of magnetic nanoparticles for their application in the field of magnetic fluid hyperthermia, *Eur J Nanomed* 5(1) (2013) 11-21. https://doi.org/10.1515/ejnm-2012-0008.

[219] Wang, X. Q., A. S. Mujumdar, A review on nanofluids - part II: experiments and applications, *Braz J Chem Eng* 25(4) (2008) 631–648. https://doi.org/10.1590/S0104-66322008000400002.

[220] Nakano, M., H. Matsuura, D. Y. Ju, T. Kumazawa, S. Kimura, Y. Uozumi, N. Tonohata, K. Koide, N. Noda, P. Bian, M. Akutsu, K. Masuyama, K. I. Makino, Drug delivery system using nano-magnetic fluid, *3rd International Conf Innovative Comput Inform Control, ICICIC'08* (2008) 8–11. https://doi.org/10.1109/ICICIC.2008.237.

[221] Scherer, C., A. M. F. Neto, Ferrofluids: Properties and Applications, *Braz J Phys* 35 (3) (2005) 718–727. https://doi.org/10.1590/S0103-97332005000400018.

[222] Hedayatnasab, Z., F., Abnisa, W.M.A.W. Daud, Review on magnetic nanoparticles for magnetic nanofluid hyperthermia application, *Mater Design* 123 (2017) 174-196. https://doi.org/10.1016/j.matdes.2017.03.036.

## Further Reading

Hedayatnasab, Z., F. Abnisa, W.M.A.W. Daud, Review on magnetic nanoparticles for magnetic nanofluid hyperthermia application, *Materials Design*, 123, 174-196 (2017). Elsevier. https://doi.org/ 10.1016/j.matdes.2017.03.036.

Ohring, M. Magnetic Properties of Materials. *Engineering Materials Science*, Academic Press, 711–746 (1995). doi:10.1016/b978-012524995-9/50038-6.

Palagummi, S., F.G. Yuan, 8 - Magnetic levitation and its application for low frequency vibration energy harvesting, Editor(s): Fuh-Gwo Yuan, *Structural Health Monitoring (SHM) in Aerospace Structures*, Woodhead Publishing and Elsevier, The Netherlands (2016). doi: 10.1016/B978-0-08-100148-6.00008-1.

Ruiz, S.M., S.F. Bou, *Preparation and characterization of magnetic nanoparticles*, Universidad Politechnica de Valencia. http://hdl.handle.net/10251/112644.

Sharma, S. K. (editor) *Complex magnetic nanostructures: synthesis, assembly and applications*. Springer, Switzerland. (2017). https://doi.org/10.1007/978-3-319-52087-2.

Shatruk, M., J. K. Clark, Magnetic Materials. in *Comprehensive Inorganic Chemistry III*, Reedijk, J.; Poeppelmeier, K. R., Eds.; Elsevier: Oxford. (2023) v. 4, 236-261.

# Index

## A

agglomeration, 15, 22, 24, 27, 31, 35, 62, 74
analysis, 21, 31, 32, 33, 57
antiferromagnetism, 2, 5, 10, 34, 67
applications, xi, 1, 2, 12, 15, 16, 17, 22, 23, 26, 29, 35, 37, 38, 39, 40, 41, 42, 43, 44, 45, 46, 47, 48, 49, 50, 53, 55, 56, 57, 58, 62, 63, 64, 65, 66, 67, 68, 69, 70, 72, 74, 75, 76, 77, 78, 79, 80

## B

biomedical, 1, 15, 16, 27, 39, 40, 45, 46, 47, 48, 50, 53, 58, 59, 61, 62, 63, 64, 69, 72, 77

## C

candidate, 22, 53, 54
characterization, xi, 31, 32, 64, 66, 68, 69, 70, 71, 72, 76, 77, 80
coatings, 36, 58, 61, 69
coercivity, 2, 11, 12, 14, 15, 17, 29, 33, 41, 50, 63, 65
colloidal, 25, 27, 31, 35, 36, 46, 57, 59, 70, 71, 74
cooling, 53, 54, 55, 56, 59, 78
co-precipitation, 19, 20, 21, 22, 25, 27, 37, 38, 39, 40, 41, 42, 43, 44, 45, 46, 47, 48, 49, 50, 69, 76, 77

## D

decomposition, 17, 19, 22, 23, 25, 46, 47, 48, 67, 70

defence sectors, 55
dependent, 22, 34, 35, 65, 68, 73
diamagnetism, 5, 6, 9
drug delivery, 1, 15, 39, 43, 45, 58, 59, 79

## E

electronics, 56

## F

ferromagnetism, 2, 5, 9, 10, 11, 12, 13, 33, 34, 64
formation, 2, 6, 12, 15, 21, 22, 27, 29, 35, 46, 67, 69

## H

heat transfer, 53, 55, 78
hematite, 1, 3, 21, 49, 63, 64, 65, 66, 74
hydrothermal, 19, 24, 25, 34, 37, 38, 41, 45, 46, 49, 68, 71, 78

## I

iron, 1, 2, 3, 8, 10, 16, 19, 20, 21, 22, 23, 27, 33, 34, 35, 37, 38, 39, 40, 41, 43, 44, 46, 47, 48, 49, 50, 53, 54, 58, 59, 61, 62, 63, 64, 66, 67, 68, 70, 71, 72, 73, 74, 75, 76, 77, 78, 79

## M

maghemite, 1, 2, 3, 20, 21, 34, 41, 46, 47, 58, 63, 65, 66, 69, 70, 72, 73
magnetic, xi, 1, 2, 5, 6, 7, 8, 9, 10, 11, 12, 13, 14, 15, 16, 17, 19, 20, 21, 22, 23, 24,

# Index

25, 26, 28, 29, 31, 32, 33, 34, 35, 36, 37, 38, 39, 40, 41, 42, 43, 44, 45, 46, 47, 48, 49, 50, 51, 53, 54, 55, 56, 57, 58, 59, 61, 62, 63, 64, 65, 66, 67, 68, 69, 70, 71, 72, 73, 74, 75, 76, 77, 78, 79, 80, 83, 84
magnetic alloy nanoparticles (MANPs), 16, 17
magnetic metal oxide nanoparticles (MMONPs), 16
magnetism, xi, 5, 9, 11, 29, 32, 33, 35, 63, 66, 67, 69
material(s), xi, 2, 3, 5, 6, 7, 8, 9, 10, 11, 14, 15, 16, 20, 21, 22, 24, 26, 32, 33, 36, 45, 47, 54, 56, 61, 63, 64, 65, 67, 68, 69, 70, 71, 72, 73, 76, 77, 80, 83, 84
measurement(s), 13, 29, 31, 57
mechanical, 17, 35, 48, 57, 58, 83
metal oxide, 16, 22, 24, 27, 62, 66, 68
metal(s), 15, 16, 19, 22, 23, 24, 26, 27, 36, 50, 53, 54, 61, 62, 66, 68, 69, 73, 74, 75, 76, 78
microchips, 56
microemulsion, 17, 25, 26, 54, 65, 71, 78
microfluidics, 32, 56, 66, 83, 84

## N

nanoparticle(s), xi, 1, 3, 8, 11, 15, 16, 17, 19, 20, 21, 22, 23, 24, 25, 26, 27, 29, 30, 31, 32, 33, 34, 35, 36, 37, 38, 39, 40, 41, 42, 43, 44, 45, 46, 47, 48, 49, 50, 51, 53, 54, 55, 56, 57, 58, 59, 61, 62, 63, 64, 65, 66, 67, 68, 69, 70, 71, 72, 73, 74, 75, 76, 77, 78, 79, 80, 84
nuclear, 54, 78
nuclear power plants, 54

## P

paramagnetism, 2, 5, 6, 8, 10, 11, 41, 48
particle(s), xi, 1, 3, 5, 6, 7, 10, 11, 12, 15, 16, 17, 20, 21, 22, 23, 24, 25, 26, 27, 28, 29, 31, 32, 33, 34, 35, 36, 37, 38, 39, 40, 41, 42, 43, 44, 45, 46, 47, 48, 49, 50, 51, 53, 56, 58, 59, 61, 64, 66, 68, 69, 70, 72, 73, 74, 75
properties, 1, 2, 3, 5, 8, 9, 11, 14, 15, 16, 17, 20, 21, 23, 29, 32, 33, 34, 35, 36, 37, 40, 43, 46, 53, 54, 55, 58, 61, 62, 63, 64, 65, 67, 68, 69, 70, 71, 72, 73, 74, 75, 76, 77, 78, 79, 80

## R

reaction, 15, 17, 19, 21, 22, 23, 24, 25, 27, 34, 45, 46, 47, 61, 73
remediation, 54

## S

sealing, 40, 57
size, 1, 3, 12, 15, 16, 20, 21, 22, 23, 24, 25, 26, 27, 29, 31, 33, 34, 36, 37, 38, 39, 40, 41, 42, 43, 44, 45, 46, 47, 48, 49, 50, 54, 58, 61, 62, 63, 65, 66, 67, 68, 69, 70, 71, 72, 73, 78
sol-gel, 19, 27, 49, 72
space, 55
stability, xi, 1, 2, 16, 17, 20, 25, 31, 35, 36, 39, 46, 47, 51, 61, 66, 74, 75
superparamagnetism, 6, 11, 13, 29, 33, 34
surface, 1, 8, 12, 21, 22, 23, 25, 26, 28, 31, 32, 33, 34, 35, 36, 40, 44, 49, 50, 54, 55, 57, 58, 61, 62, 63, 64, 66, 70, 72, 73, 74, 75, 77, 78, 84
synthesis, xi, 1, 3, 15, 16, 17, 19, 20, 21, 22, 23, 24, 25, 26, 27, 32, 37, 61, 62, 63, 64, 65, 66, 67, 68, 69, 70, 71, 72, 73, 75, 76, 77, 78, 80, 84

## T

technique(s), 26, 31, 32, 37
thermal, 8, 10, 11, 13, 17, 19, 22, 23, 25, 27, 46, 47, 48, 53, 55, 56, 58, 64, 66, 67, 70, 78, 79

# About the Authors

**Dr. V. N. Lad** is working as an Associate Professor at the Department of Chemical Engineering, Sardar Vallabhbhai National Institute of Technology – Surat, Gujarat, India. He has more than 20 years of professional experience, and guiding students for their research leading to PhD and Master's programme. He is having a PhD in Chemical Engineering. His area of research interest includes Colloids, Interfacial Engineering, Microfluidics, Magnetic Materials, Thin Films, Process Intensification, Environmentally Benevolent & Energy Efficient Process Design, Rheology of Complex Fluids, Advanced Materials and Energy Technology.

He was the co-developer-1 for the development of the course on Mechanical Operations in pedagogical framework under the *National Mission on Education through* Information and Communication Technology, MHRD (now MoE), Government of India, anchored by Indian Institute of Technology - Kharagpur.

His credentials involve 3 research projects including one International Bilateral Collaborative research project: India-Russia joint research project sponsored by Department of Science and Technology – Government of India. He is having more than 25 international papers. He is the recipient of appreciation Gandhian Young Technological Awards – 2015 received at the Rashtrapati Bhavan, New Delhi. He is the coordinator for the state of *Gujarat,* and for the Union Territory of *Dadra and Nagar Haveli* for the State Specific Plan for Technical Education in India which provides major inputs to the *National Perspective Plan for Technical Education in India.* National Perspective Plan for Technical Education is an AICTE-MHRD (now AICTE-MoE) initiative. It is a very important component of the *Long Term Planning of Technical Education,* and will improve the usefulness of technical education in the country.

He was one of the Guest Editors for an issue of the International Journal 'Separation Science and Technology,' Taylor & Francis, UK. He reviewed many papers for various reputed peer review journals. He delivered many expert lectures and invited talks on the area of his research interest, and chaired few sessions in International Conferences. He is a life member of Indian Society for Surface Science and Technology, Indian Society for Technical Education, Indian Science Congress Association, Indian Institute of Chemical Engineers, Indian Desalination Association, Electron Microscopy Society of India, and Materials Research Society of India.

**Dr. Paritosh Agnihotri** earned his Doctoral degree from Sardar Vallabhbhai National Institute of Technology - Surat, Gujarat, India; and Master's degree from Dr. B. R. Ambedkar National Institute of Technology - Jalandhar, Punjab, India. His research area is related to fabrication of PDMS based microfluidic devices, interfacial mixing, magnetic nanoparticles synthesis and separations using microfluidics. He is currently working at the Department of Chemical and Biochemical Engineering, Indrashil University, Rajpur, Gujarat, India. He has published many research articles in the field of his research. He has also served as a reviewer for several recognised peer review journals. He has also presented many papers and delivered talks at the national and international conferences and other technical events.